Teacher's Guide

Third Edition

Introductory Physical Science

Uri Haber-Schaim / Judson B. Cross / Gerald L. Abegg / John H. Dodge / James A. Walter

Prentice-Hall, Inc., Englewood Cliffs, New Jersey

Introductory Physical Science
Teacher's Guide
Third Edition

Uri Haber-Schaim / Judson B. Cross / Gerald L. Abegg / John H. Dodge / James A. Walter

© 1977 by Uri Haber-Schaim; © 1972 by Newton College of the Sacred Heart. Copyright assigned to Uri Haber-Schaim, 1974; © 1967 by Education Development Center, Inc. Published by Prentice-Hall, Inc., Englewood Cliffs, New Jersey 07632. All rights reserved. No part of this book may be reproduced in any form or by any means without permission in writing from the publisher. This edition of *Introductory Physical Science Teacher's Guide* is a revision under free licensing of the work under the same title copyrighted originally by Education Development Center, Inc. The publication does not imply approval or disapproval by the original copyright holder.

Printed in the United States of America

ISBN 0-13-503011-0

10 9 8 7 6 5 4

Contents

Foreword iv

1 Introduction 1
2 Volume and Mass 7
3 Characteristic Properties 43
4 Solubility 73
5 The Separation of Substances 99
6 Compounds 121
7 Elements 139
8 The Atomic Model of Matter 155
9 Molecular Motion 179
10 Sizes and Masses of Atoms and Molecules 203

Appendix 222

Foreword

It has been the experience of thousands of teachers that teaching this course can be a rewarding experience provided the necessary preparations are made. It is the purpose of this *Guide* to help you reach this goal. Nevertheless, the *Guide* cannot replace a teacher-training workshop devoted entirely to *IPS* in which you will do the experiments yourself, solve many of the homework problems, and become familiar with the spirit of the course. A minimum of fifteen days (five before school begins and ten spread throughout the school year) is required.

USING THE TEXT

Some sections of the textbook are designated as "Experiments" and serve as guides to your students in their work in the laboratory. Other sections lay the groundwork for new concepts, relate the results of experiments to students' understanding of the nature of matter, or serve as introductions or summaries to chapters. These are often brief, yet they are important parts of the course. Experience has shown that it is worthwhile to have many of these sections read aloud in class and discussed in detail. Students should be asked to summarize a section for the class or to present important ideas in their own words.

THE LABORATORY IN THIS COURSE

Our knowledge of physical science is the result of years of experimentation. No student can experience all the discoveries that have been made, but as far as possible we should like him to learn physical science in the laboratory. Your students' ability to understand the discoveries of others rests on their having real experiences themselves. They profit most by making their own observations and drawing their own conclusions.

In this course, the laboratory work is an integral part of the text. Some of the significant conclusions your students arrive at in the laboratory do not appear explicitly in the accompanying text. In other words, it is assumed in many cases that students have found in the laboratory facts or laws on which subsequent sections of the text are based.

The laboratory instructions provide a minimum of directions and call students' attention to the important points in an experiment by raising questions. Sometimes the answers to these questions merely require thought; at other times experimentation is needed. Your students must decide what to do. At the beginning of the course some students may feel a little insecure with this type of laboratory work. They are likely to ask whether they have the right result. You must help them to realize that nature cannot be wrong; our job is to understand it by measurement and interpretation. If a student has not measured what he set out to measure, a discussion, rather than a yes-or-no answer, is in order.

Your students will ask for answers, and will continue to ask for them if you give them. If you let your students find their own answers, they will not only learn more but gain confidence in their ability to make useful decisions. At first you may find this difficult, but if, after listening to their questions, you give a few answers such as, "How can you find out?" "Try it," "Look it up," "You have to decide," and "Are you satisfied with the data?" your students will become resourceful.

Experiments should be done at the time they are encountered in studying the text. In this way, your students are not likely to know what to expect. As they progress in the course, they learn to enjoy doing experiments whose results they do not know in advance, even though they realize that someone has faced and answered the same problem before.

Experimental data are usually collected by individuals or by pairs of students working in the laboratory. The task of collecting sufficient data is simplified by sharing the work load among the members of the class. These data are then pooled, often in the form of tables of class data, graphs, or histograms from which generalizations can be drawn. For example, suppose a student seeks to determine if the melting point of a solid substance depends on the amount of substance (Expt. 3.2). He would have to make a number of determinations requiring several days. A properly planned class experiment in one period will provide data on a dozen samples of different sizes, which can then be pooled in a "post-lab" that will help the whole class reach a conclusion. The class generalizations often lead the class back to the laboratory for further refinements or additional experiments.

In addition to simplifying the collection of data, the class effort provides a very useful forum for discussion of ideas and results. The give-and-take atmosphere is vital if students are to learn how knowledge is acquired. Through these discussions students learn from each other as well as from the teacher.

Since the course is centered around experimental work by the students, it is of the utmost importance that the specified equipment be on hand and easily accessible. Although a well-designed science laboratory always is an asset, this course can be taught successfully in a classroom with one sink, flat tables, and a reasonable amount of storage space.

Foreword

A fair fraction of a class period usually is lost in setting up equipment and taking it down. This fraction is considerably reduced if double periods can be arranged.

To assist you in planning and conducting the experiments, the *Guide* includes information on apparatus, expected duration of the experiment, necessary materials, and recommended procedure.

Your students usually should be able to do an experiment in a normal class period; however, when this is not possible, the *Guide* indicates the best point at which to break off an experiment. In many of the experiments the *Guide* also indicates the degree of precision you may reasonably expect.

Most of the experiments are designed to be performed by two students working together. In many experiments one pair of hands is not enough to carry out the necessary manipulations, but more than two students working together can lead to confusion and wasted time.

There is some advantage in individual work in the laboratory; it forces every student to come to grips with the whole experiment and prevents one of the partners from becoming a mere note-taker. On the other hand, working in pairs gives the students more confidence in their work, and they can learn from each other by discussing their data. In any case, a notebook has to be kept by each student.

PRE-LAB AND POST-LAB DISCUSSIONS

One of the most important aspects of teaching this course is conducting a discussion of an experiment before the class attempts it (a "pre-lab") and then, after the completion of the experiment, reviewing it with the class and discussing the conclusions that may be drawn from it (a "post-lab").

Students do not automatically learn something by simply doing an experiment, even though they may have obtained very good results. In order to interpret the results and realize their implications, each student must understand why he or she is doing an experiment before starting it. In the pre-lab preceding each experiment it is advisable to involve the students in the design of the experiment as much as possible. In this way they develop a better understanding of the purpose of the experiment, the procedures they will follow, and the kind of data they will have to collect. In addition, it provides an opportunity for students to exercise their imagination and ingenuity.

Raising questions and leading a class discussion during the pre-lab is an effective way to help the students to understand the experiment and any new techniques required to carry it out. (Sometimes, as in the case of safety precautions, this is not possible, and you must simply tell them how to do something.)

For example, in the pre-lab before your students do Expt. 2.3, Measuring Volume by Displacement of Water, you might show them some sand and ask how they would find the volume the sand occupies. They may suggest pouring the sand into a graduated cylinder or into a box where the dimensions of the volume occupied by the sand can be measured with a ruler. If you then ask whether the volume determined is that of the sand or of the sand plus air, the stage will be set for them to figure out a way to find the volume of the sand alone. If you give them some time to think about it, they will probably come up with a method similar to the one described in the text, or they might suggest pouring a known volume of water into a known volume of sand. By subtracting the volume of water added from the total volume of sand and water, the volume of the sand alone can be determined. While this procedure is logical, it is difficult to do, so you might ask—if other students do not—whether it would not be better to pour the sand into the water. Most students will know from everyday experience that water seeps slowly through the sand, sometimes trapping air bubbles; consequently, they will realize that the second method is better.

Some experiments, such as 2.6, 2.7, 2.8, and 5.7, are designed to familiarize the students with an instrument or technique; others such as (4.12, 6.6, 6.7, 7.4, 8.2, 8.4, 8.5, 9.11) generate data and observations that can lead to fruitful questions or analyses. Most experiments, however, are designed to help answer an important, fundamental question. Once students understand a question clearly, there is no reason why they cannot share in the excitement of designing an experiment to find an answer.

Some experiments are more quantitative in nature and require careful recording of data, drawing of graphs, and calculations of results. Examples are Expts. 3.2 and 3.4, which deal with what happens when substances freeze or boil, and the experiments on conservation of mass in Chapter 2.

In your discussion of an experiment in the pre-lab, do not give away the expected results. (However, you do not have to pretend, with any class, that they are not known.)

In most cases you should insist that your students read the instructions for the experiment given in the text and think about why and how it is to be done before you conduct the pre-lab discussion.

As we have said, many of the quantitative experiments in the course require the collection of all the data for different conditions obtained by the whole class in order to answer questions. However, even when all the students do exactly the same experiment, they usually have time to take only one set of readings. Individual results vary, and only the pooling of results of all the stations in a post-lab will lead to useful conclusions.

Perhaps the best way to pool individual results is to draw a histogram for the results of the entire class. Such a histogram shows how separately determined values cluster around a most probable value—something that is not done by individual results or by an average value calculated from the data of the whole class. How to construct histograms is described on page 15 in this *Guide*, and examples of such class histograms are given in the discussion of the results of experiments where histograms are useful.

It is sometimes valuable to have the entire class repeat an experiment when the pooling of class results does not lead to a firm conclusion. In

Foreword

such cases, class discussion of possible errors in procedure or measurement will lead to definitive results when the class tries the experiment again. It is particularly important in such circumstances not to give away the expected results; if you do, students who have come close to the expected result will lack the incentive to repeat the experiment.

SAFETY

Since the best thing to do about accidents is to prevent them, the IPS experiments are designed to minimize classroom hazards. It should be noted, however, that a potential hazard exists whenever students are working in a laboratory. The choice of experiments and the quantities of chemicals utilized have been made after careful consideration for the safety of all involved and after thorough testing in the developmental stages. The Appendix of this *Guide* includes a complete list of chemicals and the minimum standard of quality. Our experience has shown that the major source of accidents is the improper use and handling of the materials. We therefore urge you to review and practice the following general safety procedures with your class.

Be sure that you and your students follow all local and state fire and safety regulations.

Store all chemicals in a locked cabinet (preferably in a vented storeroom) and in the original containers to avoid mislabeling.

Dispense chemicals at several stations in the classroom. This will reduce the crowding and pushing that cause spillage. (When your students measure out chemicals from a common source, care should be taken that the source is not contaminated. Students should not pour unused portions back into the containers from which they obtained them.)

Encourage the use of glycerine as a lubricant for inserting glass tubing in rubber stoppers. If possible, towels should be used to protect against cuts.

Where possible, utilize plastic or unbreakable containers for dispensing materials.

Do not allow students to substitute chemicals in experiments unless you have thoroughly checked the procedure.

Use only burner fuel, which is denatured ethyl alcohol. DO NOT use ditto fluid or other liquids that contain mainly methyl alcohol.

Never dispense burner fuel from the one-gallon metal cans. Pint-size plastic bottles should be filled (in a storeroom, if possible) away from any flame, labeled plainly, and placed at each chemical-dispensing station in the classroom.

Fire extinguishers and sodium bicarbonate solution (for acid burns) should be conspicuously placed and handy in each classroom.

Insist that your students wear safety glasses whenever they are included in the list of apparatus and materials for an experiment in this *Guide*. Be sure to wear them yourself when required.

Foreword

STUDENT LABORATORY NOTEBOOKS

It is imperative that each student keep a legible and complete record in his notebook of what he does at the time he does it. The value of a good notebook is that the student can refer to it at a later time and reconstruct the experiment from the recorded data and observations.

Do not make a fetish of neatness. It will only drive your students to recording data on scraps of paper (easily lost) and copying the data into their notebooks later, a practice that should never be allowed.

Whether the notebook is written in ink or pencil is not important. It also does not need to be a work of art; neatness and arrangement need only be sufficient to allow a student's experiment to be completely reconstructed. There is nothing wrong with abbreviations and marginal notes as afterthoughts so long as the write-up is clear. We are concerned in this course with laboratory notebooks, not a formal report.

Avoid the common practice of having students put their records in order under headings like "Object of Experiment," "Apparatus," "Diagram of Apparatus," "Procedure," etc. This tends to replace thought with mechanical details. There is no reason why every student's write-up of a given experiment should have the same stereotyped form. All the details should be clear enough if the text instructions (and the illustrations and their captions) are read in conjunction with the laboratory notebooks. However, if the procedure used differs from that in the text for any reasons such as student-suggested changes (approved by you!), change in apparatus used, etc., such changes should be briefly noted by your students in their notebooks.

Encourage imaginative students who wish to vary the procedure for sound reasons or to extend the experiment in order to answer additional questions (perhaps coming back in their free time to do it).

When checking students' notebooks, you should point out errors in spelling, grammar, and sentence structure particularly if they contribute to a lack of clarity.

All measurements taken must be recorded with appropriate units and include the name of what is being measured.

Example from Expt. 2-9, The Mass of Dissolved Salt:

Mass of container + salt	13.00 g
Mass of container	11.95 g
Mass of Salt	1.05 g

Data in notebooks are obtained from one of two sources: (1) a direct measurement or (2) a calculation. If from the latter, the entry in the notebook should show where the calculated figure came from, unless it is obvious as in the example shown above. The first example below is from Expt. 3.7, The Density of Solids. For Cube A, the equation used to find density and the figures substituted in it are shown, but because it is obvious by this time how the volume was calculated, the equation and substitution for it are not shown.

Cube A

Length of side 1	1.22 cm
Length of side 2	1.25 cm
Length of side 3	1.23 cm
Volume of cube	1.88 cm³
Mass of cube	14.65 g

$$\text{Density of cube} = \frac{\text{mass}}{\text{volume}} = \frac{14.65 \text{ g}}{1.88 \text{ cm}^3} = 7.80 \text{ g/cm}^3$$

The arrangement of the data should be such that anyone else who understands the experiment can quickly reconstruct it without any doubt as to the procedure followed, measurements made, or calculations performed. While there are many arrangements that will accomplish this, three are especially well suited to this purpose:

(1) A sequential listing of the data, calculations, and observations.
(2) A table that contains these same elements, when appropriate.
(3) Answers to questions and a conclusion, if appropriate.

If the experiment is to be run only once, a listing of the steps taken, along with the data and calculations, is usually best. This method has been used in the preceding example.

If the experiment requires that the same procedures be repeated several times, arranging the data in a table is often convenient, since it makes it easy to compare the results of different runs. (The preceding example is, in fact, only one part of a three-part experiment.) Since the same procedure that is used to find the density of cube A is used to find the density of another cube, B, and of a rectangular block, it is easier to compare the results of these three parts of the experiment if the data are arranged as shown in the following table.

Expt. 3.7. The Density of Solids

	Length of Side 1 (cm)	Length of Side 2 (cm)	Length of Side 3 (cm)	Volume (cm³)	Mass (g)	Density = mass/volume (g/cm³)
Cube A	1.22	1.25	1.23	1.88	14.65	7.80
Cube B	1.24	1.27	1.21	1.90	5.15	2.71
Block C	7.61	2.53	0.58	11.2	30.57	2.73

Let your students decide in the pre-lab, or individually, the best format for recording data and calculations in each experiment.

Answers to questions asked in the text, as well as class data and class conclusions reached in the post-lab, should be included in the laboratory notebook. Recording the conclusions is essential, since they are often important later on and may not be explicitly stated in the text.

Both introductory and specific questions are asked in the text. The introductory questions, sometimes rhetorical, form a basis or a rationale for doing the experiment. Although students should give some thought to the introductory questions asked, they are not usually expected to have a definitive answer to such questions until the experiment is concluded. For example, the beginning of Expt. 3.2, Freezing and Melting, reads, "If you live in a part of the country where it snows in the winter, you know that a big pile of snow takes longer to melt than a small one. Does this mean that a big pile melts at a higher temperature?" Before doing this experiment, students probably have no evidence from which to answer this question. It is only an introduction to the purpose of the experiment.

Of course, the fact that a question is asked at the beginning of an experiment does not necessarily mean that it is rhetorical and does not require an answer. For example, in Part A of Expt. 5.1, Fractional Distillation, students are directed to examine a liquid and are asked if they can tell, "just by looking, if the liquid is a mixture." This question, of course, requires an answer, which should include the observation that prompted it. Also, there are questions that require students to make observations that are necessary to their understanding of the experiment. In such cases it is important that students answer the questions in their laboratory notebooks, chronologically, at the time they make observations. Do not allow students to make a numbered list of answers to all the questions at the end of the experiment. Insist on self-contained answers that leave no doubt as to which question is being answered.

Since the conclusions usually are based on experimental data from the entire class, class data needed to support them should also be included in the notebook. This will also stress the need for a large amount of supporting evidence before stating a generalization. The conclusion written in the notebook should be an accurate and complete statement of the generalization made by the class in the post-lab discussion.

Many students will have had no previous experience in keeping a laboratory notebook. Therefore, it is important that the first few write-ups of experiments be given immediate and careful attention. This can best be accomplished by glancing at students' notebooks, asking questions, and making suggestions while they are doing an experiment. The extent to which this can be done will depend on the experiment. For example, if a measurement is being made every 30 seconds, it would disrupt the work to question a student about his notebook. If possible, sit down with a student out of class and carefully go over an experiment, asking him questions about the entries in his notebook and making suggestions as to how he might correct the deficiencies. If this is not possible, collect the notebooks and write suggestions for improvement in the margins. Then see how well the improvements are made.

Foreword

An excellent section on laboratory reports can be found in Eric Rogers' *Physics for the Inquiring Mind*, Princeton University Press, 1960, p. 64.

THE PROBLEM OF ABSENCES

Because laboratory experiments are so frequent, a student who is absent from class is likely to miss one or more of them. Since important generalizations and conclusions are reached through the results of student experiments and often are not stated explicitly in the text, a student who has been absent is at a special disadvantage. Rather than have the student make up experiments after school, which often is not practical, you can have the student discuss in detail with his or her lab partner or another classmate the techniques and problems encountered in doing the experiment, and then copy the class data and conclusions arrived at in the post-lab session into the notebook.

TEACHER DEMONSTRATIONS

In several cases the textbook describes experiments not done by the class, giving actual data obtained with equipment shown in the illustrations. You should demonstrate as many of these experiments as possible, or ask teams of students to prepare and present the demonstrations.

Six of the demonstrations are available on film loops. When you do not have the necessary equipment or material, such as a Geiger counter or Dry Ice, the loops are extremely helpful in demonstrating the phenomena that are discussed in the text. But even when you do the demonstration, the film loops are handy for review purposes.

END-OF-SECTION AND END-OF-CHAPTER QUESTIONS AND PROBLEMS

The text contains a large selection of problems—some easy, short, confidence-building ones; some more complex; and, finally, some that go beyond the course and serve as an optional extension of the material. This *Guide* will help you to identify the various types of problems. In general, it is up to you to decide which problems will be most suitable for your class.

The problems and questions found at the ends of sections generally cover single concepts and are designed to reinforce ideas immediately after they are encountered in reading the text or in the laboratory. A selected number of these are marked with a dagger(†), indicating that the answer can be found in the appendix of the text. These will help your students decide if they understand the ideas in the sections directly preceding these "daggered" problems and are ready to go on.

The set of problems labeled "Home, Desk, and Lab" located at the end of each chapter follows the order of presentation of material within the chapter. These HDL's are designed to extend the students' knowledge to more general applications of the chapter concepts and, in some cases, provide additional practice with important ideas. In addition to questions and problems there are some short laboratory exercises. Many of the HDL's can be assigned on an individual basis depending on students' needs or abilities. Thus it is not necessary to assign the same problems to all students. In particular, you can use the harder problems as extra assignments for the faster students who wish to go deeper into the material. This is much more beneficial than letting them run ahead of the class, which interferes seriously with class discussion and class pooling of data in post-labs, in which important general conclusions are reached.

To get the maximum benefit from the HDL problems at the ends of the chapters, it is important to assign them at the time when the subjects with which they deal are being discussed in class. Call on your students to present their solutions to assigned problems and defend them before the class. Many of the questions raised in the homework problems, as well as in the experiments, have more than one answer depending on the assumptions a student may make. Do not be tempted to judge the students' answers with a simple "right" or "wrong." Instead, ask for the reasoning behind the answers. It is better to assign fewer problems and have them treated in this way than to have the class do more and hand them in, say, once a week simply to be marked. Assigning 50 to 80 percent of the problems seems to be appropriate, depending on the number and the length of class periods and the ability of your students. Since topics and techniques from the early chapters of the text are used throughout the year, we recommend that from time to time you assign one or two problems from earlier chapters that are particularly relevant to the current topic.

REVIEWING THE "STORY LINE"

As your students progress through the course, be sure to stop now and then after covering a single important chapter or several closely related chapters to review with your students the "story line"—the path of step-by-step logical development and building up of evidence over which they have traveled from the beginning. This is particularly important in this course, because succeeding chapters and topics are closely tied together by a tight, logical development.

ARITHMETIC SKILLS

The IPS course presupposes only standard facility with arithmetic. Experience in working with ratios is desirable. In addition, students will find a slide rule particularly useful for calculations. The experience of many teachers has shown that arithmetic and slide-rule skills require

Foreword

repeated use if student competency is to increase as the school year progresses. An occasional class period devoted to a review of some useful arithmetic may be necessary. However, extended mathematics sessions, particularly in the beginning of the year, are boring and dampen student interest.

The increasing availability of low-cost hand calculators makes it likely that some of your students will own them. Whether or not to permit the use of calculators in an IPS class is the teacher's decision to make. Certainly their use relieves the tedium of some of the necessary arithmetic and helps to prevent numerical errors. If you choose to permit their use, insist that the students discard nonsignificant digits (see Section 3.6). If a student wishes to lend his or her calculator to a classmate, insist that the calculator be cleared before handing it over to the person who borrows it.

JUDGING ACHIEVEMENT

Achievement in this course manifests itself in many ways, some of which are not subject to quantitative measurement. Consider, for example, the student who, at the beginning of the year, is quite lost in the laboratory and finds it hard to make a move without explicit instructions from you; yet a few months later you see this same student working independently and knowledgeably. The student has certainly progressed and achieved. Another intangible is the improvement shown in students' skill in communicating orally the results of their laboratory work or the reasoning behind the solution of a problem. Progress in these domains, although hard to express quantitatively, should certainly be considered in arriving at a student's grade.

Your "paper work" in teaching this course need not be a burden if you are selective in what you read and mark. It is unnecessary and, in fact, impossible to read and correct every detail of all questions and HDL's a student does and all the experiments in a laboratory notebook. Read only a few notebooks and homework papers at a time. When you do correct student written work, be sure to do a thorough, thoughtful, and careful job. You must be serious about the comments you make if you wish your students to respect them. Be sure to check from time to time to see if students have heeded your suggestions.

You can do much in evaluating laboratory work by checking your students' laboratory notebooks and their handling of different techniques as you move around the laboratory observing, questioning, and helping them (only when they really need help!).

In your students' work—such as laboratory notebooks, HDL solutions, and oral presentations—errors in spelling, grammar, and sentence structure should be pointed out, particularly if they interfere with clarity. However, in assigning grades it is best to give no weight to such errors *per se*, only to whatever lack of clarity results.

It is good practice in this course, in regard to those evaluations that are subjective by nature, to reward and encourage students who make progress in these areas but not to downgrade those who do not. You must remember that your students are individuals with individual differences, and their work cannot all be judged on the same basis.

The set of laboratory tests that accompany the course are designed to assist in ascertaining your students' ability to plan, organize, and resolve a laboratory problem. In addition, you may wish to utilize some of the HDL's as laboratory tests for some students.

To help you to determine your students' overall comprehension of the course, we have developed two sets of five objective achievement tests that cover the entire course and are consistent with its objectives. These tests are available from Prentice-Hall, Inc. We suggest that you use them. Students should know that the mean scores on these tests are usually about 50 percent and that they will not be graded by the usual percentage-letter grade relationship. You should also realize that these tests can be very helpful in evaluating the effectiveness of your teaching. The *Teacher's Testing Manual* for the test package provides detailed information on the diagnostic use of the tests.

It has been the experience of our pilot teachers and students that the only way to prepare for IPS tests is to work consistently in the course. Cramming is neither necessary nor useful.

Do not rely entirely on multiple-choice tests like the *IPS Achievement Tests*. Give your students an occasional HDL-type test so that you can judge their ability to organize and communicate their thoughts on paper.

We have found that it takes a long time to generate really good questions in the spirit of the IPS course. We suggest that for an essay-type test you utilize some of the HDL's in the text, augmented by one or two of your own questions. In writing your own questions, be sure that you put the emphasis on the broad topics that are stressed in the text and in the laboratory experiments. If you write questions that rely too heavily on memory and recall, your students will quickly catch on to the fact that the way to get high grades is to cram for your tests. "Open book" and "open lab" tests can be used to advantage in this course. If you feel your own questions cannot be used in "open book" tests, it shows that they need revision.

This course is definitely not suited to a straight averaging of all grades at the end of the year. Combining the averages of all five IPS multiple-choice tests, laboratory tests, essay tests, and laboratory-notebook grades (if you grade them), does not give a useful result. This is because, instead of consisting of nearly independent "units," the course develops throughout the year as a series of closely and logically connected steps. The later parts of the course are built on the foundations laid in the earlier parts in such a way that most concepts, once introduced, keep reappearing throughout the course. Thus, for example, a student who did not understand density very well at the beginning of the year should not have a low grade earned early in the year averaged with a high grade earned near the end of the year. The latter might be due, in large measure, to a clear understanding of density as it applies to the effect of pressure

Foreword

on the density of a gas (Chapter 9). Therefore, for students who have shown steady improvement throughout the year, the weight given to the grades accumulated during a marking period should be increased as the school year progresses.

In summary, we have made a list of the items that can be used in evaluating your students' achievement. They are:

>IPS Achievement Tests (multiple-choice, objective)
>IPS laboratory tests (subjective)
>HDL-type tests (subjective)
>HDL's, on occasion (subjective)
>Laboratory notebooks, on occasion (subjective)
>Communication skills (subjective)
>Laboratory and other skills (subjective)

Read the Epilogue in the student text and keep in mind what it says whenever you make a judgment of your students' achievement.

THE INDIVIDUAL IN THE CLASS

It probably has become clear from what has been said so far in this Foreward that the IPS course offers a wide variety of experiences to the individual while providing him with the benefits of interacting with the class as a whole. However, for the sake of emphasis it may be worthwhile to summarize the various ways in which you can personalize your instruction.

1. *In the Laboratory.* Once the general purpose of an experiment has been established in the pre-lab and the class gets down to work, you can divide your time among the students according to their individual needs. Spend little time with those who are well on their way. Help, with a few leading questions, those who are encountering difficulties. Most experiments are open-ended; that is, there is always something useful to do for those who have finished the minimum assignment ahead of the others. Encourage them to go on.

2. *Individual Contributions to Class Discussions.* Every class has its extroverts and its introverts. Since being able to communicate ideas is an important goal of this course, call on students selectively so as to give those who need it more opportunity to present their solutions to problems or the results of their experiments to the class.

3. *Individual Depth.* The degree and direction of the interest of the individual students in the course will vary. Permit some of them to concentrate on the concrete and qualitative and treat lightly the abstract and the quantitative. On the other hand, let others go to greater depth. Challenge them with harder HDL's. Let them help you in setting up class demonstrations and in teaching their peers.

These three ways of personalizing your relations with your students serve individual needs while making full use of the class as a learning community. We feel that this approach answers the needs of the individual far better than letting individual students go at their own pace through the *same* material to the *same* depth.

SCHEDULING THE COURSE

The most appropriate speed with which to proceed in the course varies over a wide range and depends on the ability of the students, the size of the class, and the number and length of class periods. In a large class of below-average students you may find it profitable to go very slowly, spreading the first eight chapters over the entire year, perhaps treating the end of the course very lightly; whereas with a smaller and more talented group you may find it more challenging to proceed at a greater speed with the first six chapters, going into greater depth in the second half of the course. Therefore, the overall schedule suggested here (it does not include the laboratory tests) should be regarded only as a very general framework.

Chapter	Periods	Chapter	Periods
1	3-4	6	16
2	19	7	11
3	14	8	15
4	15	9	12
5	15	10	15

A more detailed schedule appears at the beginning of each chapter in this *Guide* and in the laboratory test package. These detailed schedules should also serve for general orientation only.

Uri Haber-Schaim
Judson B. Cross
Gerald L. Abegg
John H. Dodge
James A. Walter

Introduction

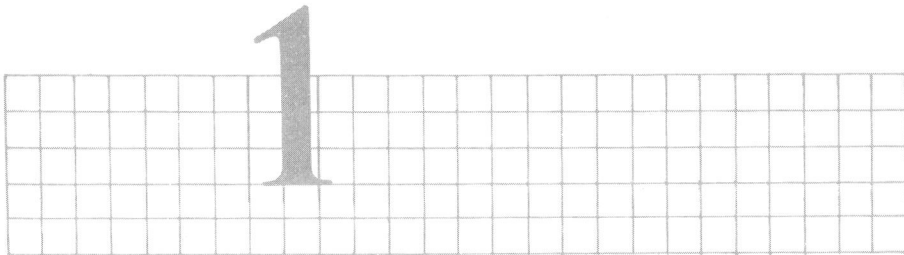

In this course careful reading is essential. A good way to make your students aware of this is to read the introduction to the book with them in class. In particular, spend enough time discussing the questions raised in the middle part.

You are likely to find two kinds of students in your class. One kind will never have thought about these questions before, and many of the terms will be entirely new to these students. The other kind will have "all the answers," using scientific terms such as atoms, molecules, and electrons without having an understanding of where they come from or what they mean. If you get such statements from students as "matter is made of atoms, and atoms are made of neutrons, protons, and electrons," this is your opportunity to show how science is to be studied in this course. Ask the student who gives you this type of answer how he knows that there are atoms, how he knows that there are electrons, what he knows about their behavior, etc. Experience has shown that a thorough discussion of this kind early in the course contributes significantly to convincing your students that during this course they will be expected to justify and defend their answers, not just parrot what they have memorized.

The first experiment is intended to raise questions, but—even more important—it gets your students out of their chairs and working in the lab on a meaningful experiment. The questions raised in the introduction should be covered well enough to give some valid reasons for going to the laboratory, but do not spend too long on them. A good way to move into the laboratory quickly (even on the first day is possible) is to have your students assemble and disassemble the apparatus for Expt. 1.1. You can quickly demonstrate how to get started, and if you name the parts as you place them on the pegboard, students will learn what they are in a natural way.

EXPERIMENT: DISTILLATION OF WOOD   1.1

To find out about matter, you must do things to it—in other words, perform experiments. Students learn from this first attempt that experiments raise questions and suggest further experiments. They also become aware of the need to find a good way to measure the quantity of matter, which will be studied in the next chapter. The purpose is not for students to identify particular substances that they produce or to try to determine the true composition of wood. Nor can they settle the question whether the substances found were present as such in the original wood or have been produced by heating. The importance of the experiment lies in the fact that questions are raised that will be discussed as the course progresses.

The Experiment

This experiment introduces your students to the laboratory. They will not be familiar with the equipment, and it will probably take the better part of a class period just to identify and assemble the necessary apparatus. Since a variety of difficulties may arise, it is helpful for your students to see a completely assembled apparatus; this also will develop their powers of observation, more than being told or shown.

The glass tubing may break when inserted in the stoppers, and so it is well to have an ample supply of glass bends. To reduce the danger of breakage, students should always lubricate the end of the glass tubing with glycerin and should grasp the tubing near the lubricated end while pushing it into the stopper. Towels or pot holders held around the tubing and the stopper as the tubing is inserted will prevent serious accidents. For the distillation of the liquid, a one-hole No. 2 stopper with a short right-angle bend is convenient; this item is used in other experiments and might well be kept assembled and available.

If possible, let each group set up the apparatus for itself. In most experiments, it is wise to check each experimental setup before your students proceed. Manipulating the equipment, maintaining a notebook, and making critical observations are, very likely, new experiences for your students. As they become more proficient in these tasks, they will gain confidence in their ability to work by themselves with a minimum amount of direction or interference.

Since the experiment will take more than one laboratory period, you will have to decide on a convenient stopping place that will not entail doing part of the experiment over again the next day.

A good way to organize the experiment is to spend the first day familiarizing your students with the apparatus. During the second day the apparatus can be quickly assembled, the wood can be heated, and the gas and liquid products can be collected, The third day can be used to distill the liquid product and to summarize the observations in a post-lab.

The pegboards and clamps are designed so that the assembled test tubes can be easily removed and stored without disturbing the rest of the apparatus, which can be left in place for the next class. If more than one class will use the apparatus, an extra supply of test tubes, stoppers, etc., will be needed. Also, provision must be made for students to label and store the intermediate products of the distillation. Masking tape and marking pencils can be used to label the test tubes and the test tubes themselves may be stored upright in a large beaker or clamped to an unused pegboard.

If your alcohol burners do not have wick caps to reduce evaporation, fill them two-thirds full on the day of the experiment. If they are filled the day before, much of the fuel will evaporate. The top of the wick should not extend more than about 3 mm from the end of the brass sleeve; otherwise the wick will burn as well as the alcohol. For the same reason, the burners should not be used when the alcohol level is very low. A filling of alcohol should last about an hour.

The water in the beaker containing the condensing test tube should be as cold as possible in order to provide effective cooling.

When boiling chips are employed to prevent violent boiling and spattering, new ones should be used for each boiling operation. Only two or three very small chips are needed. Old, used chips can be reactivated by drying and heating, but it may be more efficent to throw them away.

Since the amount of product obtained will depend on the amount of wood used, have your students pack as many splints as possible into the test tube, but make sure that enough space is left so the rubber stopper will not be burned when the splints are heated.

A short time after the wood begins to heat, water vapor condenses on the walls of the distilling tube. Approximately 4 min after the beginning of the experiment, smoke and gas appear in the tube. This smoke and gas will flow into the condensing tube, where part of it is collected.

Warn your students to remove the condensing tube before they stop heating. This prevents cold liquid from backing up into the hot test tube.

Once the bottle is full and the rubber tube is disconnected, the gas coming out from the condensing tube is relighted so that as the heating is continued the gas will not fill the room. The gas is collected only to give the students some idea of the large volume of gas produced.

Heating should continue for about 10 more minutes. This will allow at least 5 cm^3 of liquid to collect in the condensing tube. During this heating your students should remove the burners from the clamps and move them along the entire length of the tube to ensure that all the wood has been strongly heated. As said earlier, the experiment can be interrupted here and the condensed liquid allowed to stand overnight. Before redistilling the liquid, the rubber or plastic tubing used should be cleaned out with alcohol or burner fuel.

In redistilling it may be necessary to remove the burner stand and heat the test tube containing the condensed liquid directly. The tubing should again be rinsed out with alcohol right away to get rid of any contamination.

Before class discussion of the results, cleaning up must be done. This should always be done by your students. Have everyone in the class responsible for his or her own apparatus and see to it that this responsibility is assumed immediately.

The questions that are asked in the text about the experiment must be answered by each student and kept for reference in his or her notebook.

Answers to Questions

Students may predict a variety of possible results before heating the wood, such as burning, smoke, etc. The purpose of this prediction is to point out to the students that only by doing the experiment, by performing operations on matter, can they really be sure of what happens.

When the wood gets hot enough, liquid begins to condense on the walls of the test tube. (The students do not know whether it is water or some other liquid, or a mixture of liquids.) Smoke flows into the condensing test tube, and liquid begins to condense in this tube.

The gas that comes out of the apparatus burns with a blue-and-yellow flame. The condensed liquid may appear at first to be only one liquid, but if allowed to stand, it separates into two layers, an almost colorless liquid on top of a dark-brown liquid.

When the liquids are distilled, the brown liquid is left behind and the colorless or light-yellow one is distilled off and condensed in the right-hand test tube. (If the distillation is carried on too long, the condensate will become yellow to brown.)

When the two liquids are mixed together, the original brown mixture is obtained again if the mixture is thoroughly shaken.

The charcoal remains of the wood splints glow red in the flame of a match or burner and gradually disappear. They "burn," but if the wood was heated sufficiently in the first place, they burn with a nearly invisible flame—like charcoal briquets in a broiler—and a small amount of white ash remains.

You could not predict the results of this experiment without having performed it or a similar experiment. You cannot reconstruct the wood by mixing the products of the distillation. (This point will have very important implications in future chapters. Here it should be discussed only in relation to the experiment.)

It is impossible to decide whether the substances obtained were originally present in the wood or were produced by the heating. There is no evidence to help answer this question.

Different substances were obtained because heating the wood either separated substances originally present or resulted in the formation of new ones.

The amounts of solid, liquid, and gas are most easily compared by volume, roughly by just looking at how much of the various containers are filled by the substances and by estimating the volume of gas that was burned and not collected. The solids and liquids could be massed, but the students probably would not know how to mass a gas. The process of massing should not be stressed. It is the subject of the next chapter.

Apparatus and Materials

Pegboard
4 Clamps
3 Small test tubes
Beaker (600 cm^3)
Quart bottle to collect gas
Plastic bucket
2 Glass bends (5-mm tubing)
No. 2 one-hole rubber stopper
No. 2 two-hole rubber stopper
10-in. piece of rubber or plastic tubing

2 Alcohol burners and fuel
Burner stand
Boiling chips
Pine-wood splints
Matches
Cold water
Marking pencils
Glycerin
Safety glasses

The test tube containing the wood and the one containing the distilled liquids can be cleaned by soaking them overnight in a good lab cleaner like Alconox or Soilax. The tubes that held the wood may not clean very well, but they can be used again for the same experiment.

The rubber tubing should be cleaned out with alcohol.

Volume and Mass

An Overview of the Chapter

In daily life, volume and mass are often used interchangeably to measure quantity of matter. Experience has shown that the two concepts are not well separated in the students' minds.

We begin with volume, showing different methods of measuring it and the need to say precisely what we mean by the volume of an object. Through the shortcomings of volume as a measure of the quantity of matter, we then proceed to mass, which is operationally defined as that property of matter that can be measured with an equal-arm balance.

The balance introduced in this chapter will be used throughout the course. This is the time for your students to learn to handle it properly.

The experimental work in this chapter is aimed primarily at showing the advantages of mass over volume as the measure for the quantity of matter, leading to one of the most fundamental laws of nature—the conservation of mass. The suggested schedule for this chapter is:

Sections 1-5 (one experiment, problems 1-13; 26-35; 44-45)	3 periods
Sections 6-8 (three experiments, problems 14-18; 36-41)	6 periods
Sections 9-15 (five experiments, problems 19-25; 42-43)	8 periods
Achievement Test No. 1	2 periods
Total	19 periods

VOLUME 2.1

The process of measuring length, volume, or any other quantity consists of counting units. In measuring the length of an object one counts the number of length units that make up the length of the object. The measurement of volume is introduced in the same manner by counting the number of unit volumes that will fit in the unknown volume. The text and probs. 1, 2 and 27 are designed to develop this method of finding the volume of an object. Only after the basic idea of counting unit volumes is well understood do we proceed to the formula, volume = length × width × height, when the object is rectangular, as a "shorthand" way

7

of counting cubes. This is designed to prevent rote memorization of the formula without understanding its meaning.

The particular units that are counted are quite arbitrary. Only the metric system is used in this course, and it is introduced slowly. Experience has shown that a gradual introduction without tedious conversion exercises is one of the most effective methods of teaching the metric system. You should insist that all data be reported and recorded in metric units.

It is worthwhile to make sure that your students realize that volume, area, and length are different kinds of quantities and cannot be compared; for example, 5 cm is not equal to or more or less than 10 cm^2 or 2 cm^3.

Measuring the volume of an irregular object by the displacement of water is an excellent demonstration or experiment that will be very helpful in preparing the students for the next experiment. See the note in the next section of this guide concerning the use of graduated cylinders.

2.2 READING SCALES

In some of the experiments to come, the key to valid conclusions will be the reading of a scale to the highest accuracy possible. The purpose of this section is to introduce your students to estimating fractions of a scale division to the nearest tenth.

There is a special problem in reading the level of a liquid in a graduated cylinder because of the formation of a meniscus. Before you begin the next experiment, discuss prob. 6 with your students. After the discussion, have them read the liquid level shown in Fig. D; they should see that the best reading is 22.0 cm^3.

2.3 EXPERIMENT: MEASURING VOLUME BY DISPLACEMENT OF WATER

To demonstrate measuring volumes of irregularly shaped objects, students use a graduated cylinder to measure the volume of a sample of sand, first when the sand is dry and then by water displacement. The purposes of the experiment are (1) to increase students' understanding of what is meant by volume, (2) to show how it is possible to measure the volume of an irregularly shaped object by displacement of water, and (3) to emphasize that with certain materials, like sand, we must specify how the volume was measured. Volume has its limitations as a measure for matter, and this experiment will help lay the groundwork for later discussion leading to a consideration of mass as a better measure.

Dry sand with particles in the range of 2 to 4 mm is better than very fine sand, which has a tendency to pack and cause some difficulty when the time comes to remove it. After the dry sand is poured into the dry graduate, tapping the graduate gently will cause the sand to settle a small amount. No attempt should be made to pack the sand in any other way; whether or not the sand is well packed is not important. Be sure that you provide a container for disposing of wet sand. Putting it into the sink may cause expensive plumbing problems.

If the water is added to the sand, and not sand to water as directed in the text, there may be inaccuracies due to air pockets. If dry graduated cylinders are not available owing to consecutive classes, there will be some difficulty in measuring the volume of sand poured into the graduate because some grains will adhere to the side of the cylinder. This difficulty can be avoided by reserving a number of graduated cylinders to be used exclusively for the measurement of the volume of dry sand, or by carefully drying the cylinder with a paper towel. An alternate solution is to use small pebbles instead of sand. Wet sand can be removed from a graduated cylinder by repeatedly stirring the sand with excess water and pouring this slush into the disposal container.

If you give each group a different amount of sand, the class will be able to conclude whether the fraction of air space in the sand depends on the amount of sand used. This technique of assigning different amounts of material to each group will be used repeatedly throughout the course.

Sample Data And Answers to Questions

The results given below are typical:

Volume of dry sand	36.0 cm^3
Volume of water placed in the graduated cylinder	18.3 cm^3
Volume of sand plus water	39.4 cm^3
Volume of sand alone, measured by water displacement	39.4 cm^3 − 18.3 cm^3 = 21.1 cm^3
Volume of air space in 36.0 cm^3 of dry sand	36.0 cm^3 − 21.1 cm^3 = 14.9 cm^3
Fraction of air space in the sand is	14.9 cm^3/36.0 cm^3 = 0.41 or 41%*

*A clumsy fraction like 14.9/36.0 is best expressed as a decimal.

Apparatus and Materials

Graduated cylinder (50 cm^3)
Sand, dry (about 40 cm^3)
Beaker (250 cm^3)
Water
Several buckets or other containers for collecting the wet sand

SHORTCOMINGS OF VOLUME AS A MEASURE OF MATTER 2.4

For the demonstration shown in Fig. 2.7 you will need rock salt and you can use the long, straight glass tubes furnished with the IPS equipment. These are 6 mm OD (outside diameter) and 4 mm ID (inside diameter). Setting this up is a good student project.

The narrow tubes leading into the wide test tube are the first amplifiers that the students encounter in this course: A small change in volume will register a considerable change in the level of the water. We suggest you call their attention to this.

2.5 MASS

There is a great deal of confusion about the meanings of mass and weight, but a lengthy discussion here cannot be justified. The course does not need it, and students are not prepared for such a discussion. The weight of an object on earth is the pull the earth exerts on it. Weight can be measured by hanging an object on a spring. The greater the object's weight, the more it will stretch the spring. On the moon it is the moon's pull. Objects weigh one-sixth as much on the moon as they do on earth. The quantity measured with an equal-arm balance is mass. The mass of an object is the same everywhere—on earth, on the moon, or in outer space. At any given location the weight of an object is proportional to its mass. Locally we can use a balance (equal-arm or spring) to measure either quantity because at the same place objects of equal mass have equal weight. This is the reason for the statement that a grocer's scale reads 2.2 pounds when you place a mass of 1 kg on it.

An unknown mass is always compared with a known mass on an equal-arm balance. We could do the same with a spring balance. However, the deflection of a spring balance made by a given mass on the earth will not be the same as on the moon.

The best way to stay clear of the confusion between mass and weight is to use the term *mass* exclusively in this course.

2.6 EXPERIMENT: THE EQUAL-ARM BALANCE

The purpose of this experiment and the next two is to acquaint your students with the equal-arm balance. If you allow plenty of time for these experiments and cover them very carefully, the important experiments on the conservation of mass will proceed more smoothly. Since your students will use the balance frequently during the remainder of the course, skill in its use is very important. The time spent here in making sure that the balances are working properly and in helping students learn how to use them will save considerable time later.

The pointed metal rider on the left arm of the balance, used in adjusting the balance to the zero reading, should fit tightly enough on the arm to stay in place but be loose enough to move for adjusting. If the balance does not adjust to the zero reading with the left-hand rider between one-third and two-thirds of the way along the left-hand arm, try interchanging the balance pans. If this does not correct the condition, an additional mass may have to be added to one arm (masking tape, rubber band) in order to get a zero balance point. Allow enough time to get all balances properly adjusted.

One should observe several swings of the pointer. Each successive swing will be a little smaller than the previous one; but the pointer should make about ten complete swings before coming to a stop, unless the agate V bearings are dirty or the pointer is rubbing on the central support.

If the balance still does not swing freely, check to see that the load is placed on the pans so that the pan support rods are vertical and not binding at the end slots of the balance arm.

Next, if possible, check to see if the end and center fulcrum rods are straight and all parallel to one another by viewing the balance from both ends. Also check the knife edge which rests on the agates, to see that it is smooth and sharp.

All readings of the balance should be made when it is swinging freely. To wait for it to come to rest not only is time-consuming but also may be inaccurate, since friction may finally bring it to rest at a position that is not the true balance position.

Whenever the balance is to be used, the zero setting should be checked, since handling of the balance and slightly off-level table tops will change the balance pointer's zero position.

The Experiment

Once the balance is in adjustment, your students are ready to find the mass of various objects to the nearest 0.1 g. You can provide objects such as test tubes, small plastic containers, coins, paper clips, etc., to be massed. It is unwise to mass objects above 50 g. The balance can support masses greater than this, but its performance is impaired.

In using the balance for the first time, your students will find that their balances will not mass to less than 0.1 g, the smallest standard mass in the set. To mass objects to a smaller fraction of a gram they need to calibrate their rider scales. This is part of the next experiment. In the present experiment the goal is to mass objects only to the nearest 0.1 g, to give students facility in using a simple balance.

A typical set of massings recorded by a student might be the following:

Test tube	more than 16.0 but less than 16.1 g
Pencil	more than 4.6 but less than 4.7 g
Paper clip	more than 0.5 but less than 0.6 g

By massing only to the nearest 0.1 g, students will become familiar with their balances without the added confusion of the rider scale. At the same time they will discover that the balance itself is sensitive enough to mass to less than 0.1 g. This sets the stage for the introduction of the rider scale in the next experiment.

Apparatus and Materials

Equal-arm balance
Set of gram masses
Various objects such as test tube, paper clip, filter paper

EXPERIMENT: CALIBRATING THE BALANCE 2.7

In this experiment, students measure the masses of the pennies only to the nearest 0.1 gram. The rider on the right arm of the balance is kept as near the center of the balance as possible, even if a previous class has left calibration marks on the balance arms. The coins are unlikely to differ in mass by as much as 0.1 g.

The Experiment

Calibrating the balance in 0.01 parts of a gram can be done in ways other than that described, but none of these are necessarily simpler. If you have more than one class using the apparatus, putting pencil marks at intervals of 0.01 g on the arm itself need be done only by your first class. Students in following classes can simply check the scale for accuracy. A pencil is recommended so that any recalibration of the rider scale can be accomplished by simply erasing the old marks and making new ones.

Sample Data

Massing to the nearest 0.1 g:

Penny No.	Mass (g) between
1	3.0 and 3.1
2	3.0 and 3.1
3	3.0 and 3.1
4	3.0 and 3.1

Answers to Questions

The pennies all look the same except for old, badly worn ones, which, one would suspect, have masses slightly less than new ones.

When each penny is massed on the balance to the nearest 0.1 g, one cannot detect any difference in mass.

To mass to fractions of a gram one could use smaller standard masses.

One way to check to see if the 0.01 g marks on the rider scale are correct is to cut a long, very narrow, rectangular strip of graph paper (or a piece of wire) to a mass of 0.1 g; then cut it into ten equal pieces and place these pieces one, two, three, etc., at a time on the left pan to check the accuracy of the rider scale. (In this method your students should realize that they are assuming that the paper is cut accurately and is uniformly thick. Encourage your students to always consider what assumptions they make when drawing conclusions.)

Apparatus and Materials

Equal-arm balance
4 Pennies
A set of Class C gram masses
Scissors or wire cutters
Graph paper or thin wire

EXPERIMENT: THE PRECISION OF THE BALANCE 2.8

In this experiment students learn how to measure masses to a small fraction of a gram, and by massing two objects several times, they determine the precision of their balances. The experiment is designed to demonstrate the limits of the precision of the balance when using the same masses in different massings of light and heavy objects.

Before the rider was introduced, the mass of the penny was determined to the nearest 0.1 gram. Using the rider, the students can determine the mass of each penny to 0.01 gram or better, and it is at this point that differences in the masses of the pennies show up.

The Experiment

A penny and a rubber stopper are two objects with enough difference in mass to be used in the last part of the experiment. Students should be encouraged to ignore their previous massings of the same object so that each determination is truly independent.

It is very important for your students to understand, as a result of the data they get in the last part of the experiment, that the balances they use give values for masses that are good to at least 0.01 g. Your students can estimate between divisions on the rider to the nearest 0.001 g, since the divisions on the arm of the balance are wider than on the scales discussed in Sec. 2.2. In subsequent calculations, they can often take into account that the thousandth of a gram estimated digit is significant, thought not exact, if the balance is in good condition and if they use the same masses for several massings, adjusting only the rider.

Whether or not it pays to estimate to the nearest 0.001 g depends on the particular experiment being done. Sometimes the measured mass is to be divided (or multiplied) by another measured number of fewer significant digits. Sometimes the procedure used will itself introduce an error larger than a fraction of a centigram. In either case, the additional information obtained by estimating the final digit of the mass measurement will not be useful.

In this guide, we have given sample data only to the nearest centigram, even though in some cases, with very careful laboratory technique, it would be more accurately recorded to the nearest milligram. Do not burden your students with an overemphasis of exactly how accurately each piece of data should be recorded. Considered judgment by your students, coupled with some common sense, is enough.

If different masses are selected from the mass set to mass the same object, the errors in the standard masses must be considered. Table A gives the tolerances of standard masses conforming to class C standards, which are the masses that should be used with the IPS balances. (The mass sets should include only masses from 100 mg to 50 g.)

The rider of the old IPS "bead" balance has a mass that is slightly greater than 200 mg. If your students are using the old balances, and if you do not replace this rider, your students can calibrate the arm into 20 units of 0.01 g each.

14 Volume and Mass

Table A Class C Mass Tolerances

Mass (g)	Tolerance (mg)
50	±20
20	±10
10	±7
5	±5
2	±3
1	±2
0.5	±1.5
0.2	±0.7
0.1	±0.5

Sample Data

Massing to 0.1 division on the rider scale:

Penny No.	Mass (g)
1	3.089
2	3.072
3	3.067
4	3.087

Alternate massing of light and heavy objects:

Massing No.	Mass of Penny (g)	Mass of Stopper (g)
1	3.074	23.944
2	3.072	23.947
3	3.072	23.944
4	3.073	23.946

Answers to Questions

The pennies differ slightly in mass.

The precision of the balance has been increased, through the use of a rider, to better than 0.01 g.

Massing light and heavy objects, alternately, to 0.001 g shows that the thousandth-gram digit is significant. The balance appears to be accurate to about ±0.001 g for light objects and to about ±0.002 g for heavy objects.

Apparatus and Materials

Equal-arm balance
A set of class C gram masses
Rubber stopper
The same pennies used in Expt. 2.7

General Comments on Conservation-of-Mass Experiments

With the next experiment your students begin to accumulate evidence for the conservation of mass. Before discussing the class data for this and the other experiments on mass conservation, list on the chalkboard the original mass, final mass, and change in mass observed by each group. Then you can compare the class results by drawing a histogram or bar graph as shown in Fig. II. If you have fewer than 20 students, you may wish to have each group do the experiment twice to increase the number of readings so that you have a statistically meaningful histogram.

A comparison of the absolute change in mass is justified in these experiments if all groups use samples of roughly the same mass. In Expt. 2.10, The Mass of Ice and Water, it is likely that the mass of ice used by different groups will vary significantly. A fractional or percentage change in the mass of ice is to be preferred for plotting the histogram, because if there were a real change in mass in the process, one would expect it to be proportional to the mass of ice used.

Note that in all the experiments on conservation, with the exception of The Mass of Ice and Water, comparison of the mass before and after is made with the total mass of the container plus active ingredients. If you have a class weak in mathematics you may find it best not to raise this point at all and to plot absolute changes.

At first, your students will have some difficulty in making the distinction between an error in measurement and a change in mass. Do not try to make them think there is no change in mass if their data indicate that there is. If a histogram gives so wide a spread of results that no valid conclusion can be drawn, you can have your class repeat the experiment more carefully after a class discussion of possible sources of error, and draw a new histogram of the new results.

It should be made clear to students that only one experiment, involving only one kind of change such as dissolving salt, is not in itself very convincing evidence for concluding that mass does not change when other changes take place. This is why five separate mass-conservation experiments, all involving different kinds of change, are included in this chapter. Do not skip any; let your students do all of them in order to convince themselves of the plausibility of conservation of mass.

Histograms

In a large number of the experiments in *Introductory Physical Science*, histograms of class data, as well as graphs, are utilized. Considerable time can be saved if a large permanent grid for histograms and graphs is available in the classroom. The grid can be painted on a chalkboard or wall, or a special wall chart grid can be purchased. Some teachers have utilized an overhead projector with a transparency grid for each experiment.

A timesaving grid for histograms can be constructed utilizing a moderate-size pegboard (4 feet X 4 feet) with multicolored golf tees for data points. The axes can be made with tape (paper or masking) and labeled with a felt-tip pen. The tapes can be easily changed for subsequent experiments or they can be chalked over. The pegboard with some spacers behind it can be permanently attached to the wall, or it can stand in the chalk tray.

In drawing histograms some thought often must be given to choosing the size of the intervals used in the plot. Consider the hypothetical conservation-of-mass data in Table B. This was first histogrammed using intervals of 0.001 g as shown in Fig. I. Then for Fig. II intervals of 0.01 g were used (that is, if a value lay between 0.005 g and 0.015 g it was plotted as 0.01 g, or a value between 0.035 g and 0.045 g was plotted as 0.04 g). Finally, the histogram in Fig. III was plotted using intervals of 0.1 g.

Table B

Change in Mass
(g)

−0.144	−0.040	−0.037
+0.089	−0.024	−0.144
−0.131	−0.027	−0.124
−0.012	+0.076	−0.020
−0.010	+0.067	−0.042
+0.067	−0.010	−0.028
−0.034	−0.042	−0.041
−0.044	−0.038	−0.032
−0.040	−0.034	−0.038
−0.029	−0.041	−0.043
−0.033	−0.043	−0.037
−0.044	−0.032	

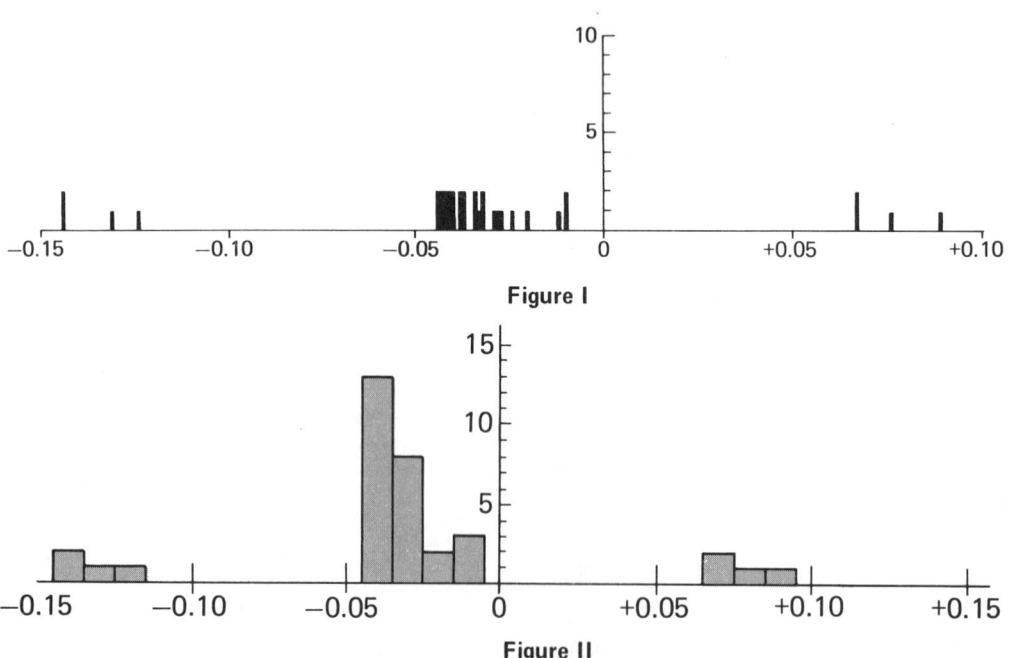

Figure I

Figure II

As you can see, both Fig. I and Fig. II show the same general features, namely that mass appears to have been lost. In Fig. III, however, the intervals are so large that the histogram is misleading. It is symmetrical about zero and suggests that there is no question: mass was conserved. Since Fig. II is easier and less time-consuming to draw than Fig. I, it represents the best choice of intervals for the data in Table B.

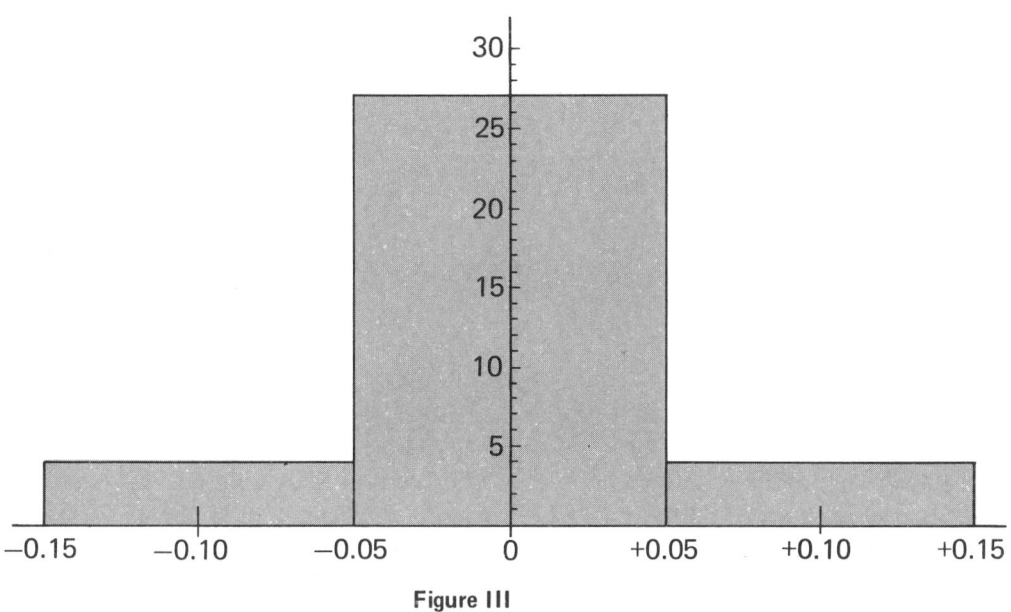

Figure III

Usually, one finds a few data points that lie exactly midway between the centers of the intervals chosen. In such cases it is best to round these up and down alternately as you histogram the data.

Be sure, when you make a histogram, that each bar you draw is centered over the value of the midpoint of the interval. This will eliminate any ambiguity as to which of two adjacent intervals the bar represents.

In most cases the origin should be included on a histogram. If it is not included, the histogram may give an exaggerated picture of the spread of the data to many students. (See Figs. IV and V.)

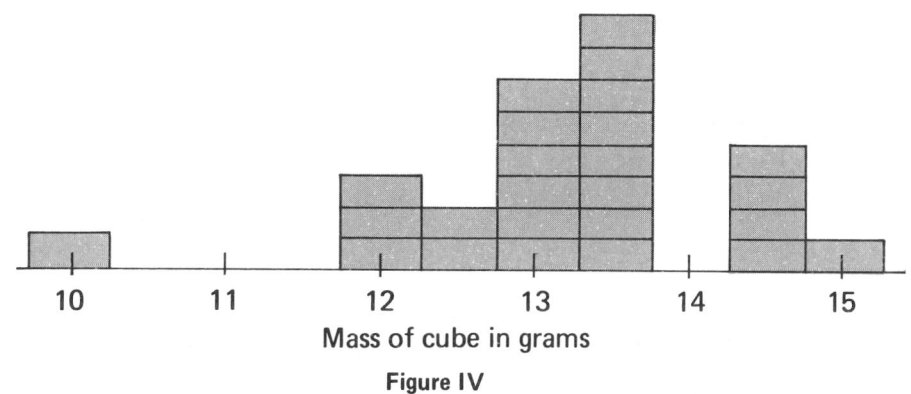

Figure IV

18   Volume and Mass

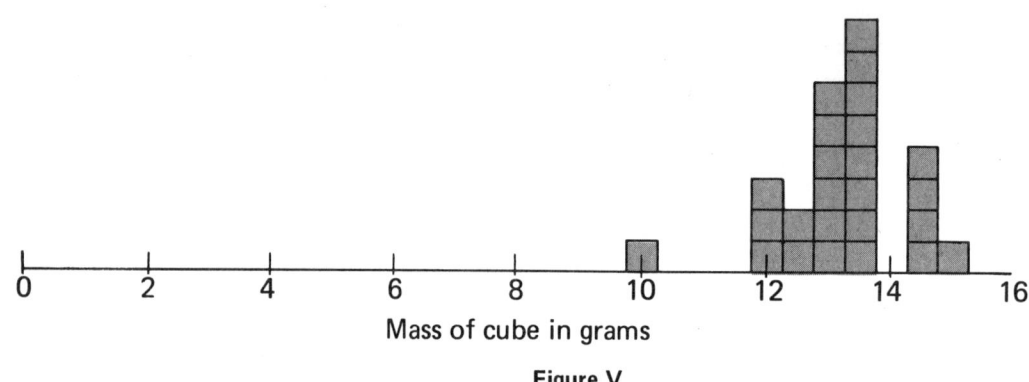

Figure V

2.9   EXPERIMENT: THE MASS OF DISSOLVED SALT

In Fig. 2.7 your students observed that a volume change took place when salt dissolved in water. This raises the question of whether there is also a mass change during the dissolving. Some students are sure that there is such a decrease and will not trust results that do not show it.

Insist that your students be very careful in making all the massings and in handling the materials. To keep the salt from spilling out of the bottle cap, a small massing "cup" of creased paper can be used in massing the salt and in transferring it to the bottle. Students may need to be reminded to include the mass of this massing cup in the final figures.

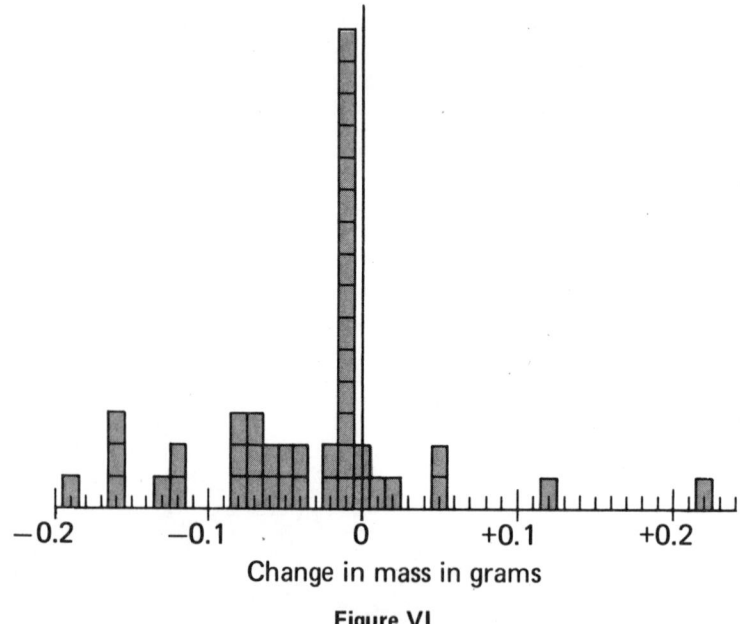

Figure VI

In the post-lab ask your class to suggest possible sources of experimental error and if the resulting errors would cause an apparent loss or gain in mass.

Do not force your students to accept experimental errors as the cause for the negative change. From the histogram in Fig. VI, for example, one is forced to conclude that there is a small loss in mass, about 0.1 g. Your students' conclusion, at this time, may be that in this experiment mass does change very slightly. In Sec. 2.15 it will be explained that similar experiments have been performed and that if any change in mass does occur, it is less than one part in a billion, the sensitivity of the best balances.

Sample Class Data

Mass before Mixing (g)	Mass after Mixing (g)	Change in Mass (g)
32.17	32.24	+0.07
29.57	28.93	−0.64
28.67	28.67	0.00
36.02	36.02	0.00
27.67	27.64	−0.03
31.55	31.55	0.00
31.45	31.43	−0.02
35.12	35.12	0.00
31.45	31.45	0.00
28.69	28.67	−0.02
31.31	31.26	−0.05
32.55	32.38	−0.17
32.26	32.21	−0.05
34.29	34.26	−0.03
31.43	31.29	−0.14
31.86	31.86	0.00
32.21	32.10	−0.11
31.40	31.26	−0.14
32.86	32.74	−0.12
31.02	31.02	0.00
30.86	30.86	0.00
30.29	30.29	0.00
28.50	28.50	0.00
29.14	29.14	0.00
30.86	30.64	−0.22

Figure VII is a histogram of the sample class data obtained for Experiment 2.9.

Change in mass in grams

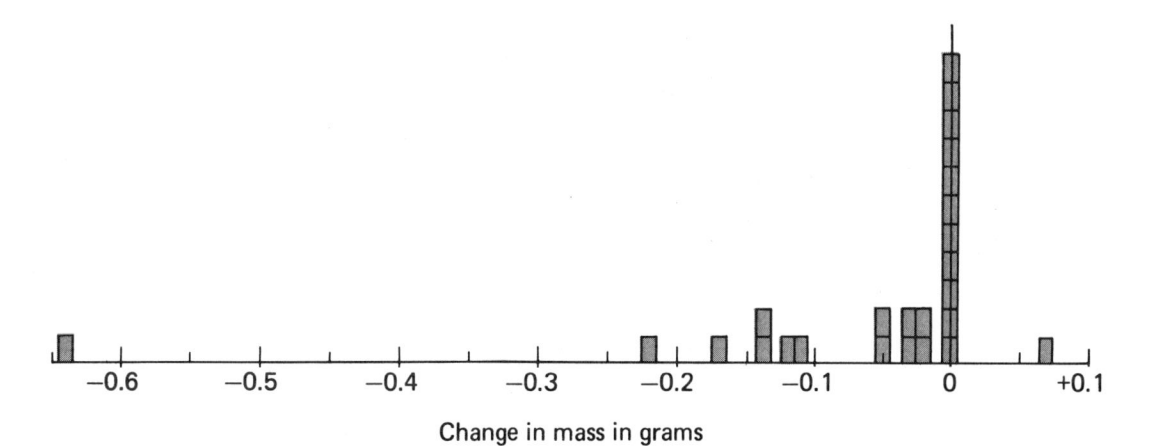

Figure VII

20 Volume and Mass

Answers to Questions

The histogram of the data (Fig. VII) shows that a zero change in mass is the most frequently obtained result, and 13 of the results are zero within 0.025 g. One result (−0.64 g) is so far from the rest that spillage or miscounting of gram masses is almost a certainty. Nevertheless, among the remaining 11 there is an indication of a loss in mass. However, the large fraction of the class that obtained no change in mass, and the lack of agreement of the others about how much mass was lost or gained, strongly supports the conservation of mass.

Apparatus and Materials

Balance
Small plastic bottle with cover
About 2 g of salt
Water

2.10 EXPERIMENT: THE MASS OF ICE AND WATER

It will take some planning on your part to complete this experiment in one class period. To speed up the melting, your students can hold the container in their hands. Expt. 2.9 or 2.11 may conveniently be performed while the ice melts.

Make sure that in each case the ice cube is small enough to fit inside the container and still allow the cover to be put in place. Your students can partially melt a cube under warm running water until it is small enough to fit easily inside the container. The plastic bottle, with the ice in it and the cover on, should be wiped dry before the first massing is made.

During the melting, the lid on the container prevents atmospheric moisture from condensing on the inside. Let your students decide for themselves what to do about the condensation that may form on the outside of the container. If they do not wipe it off, it will, of course, add to the mass of the container and melted ice. You can bring this point up in the post-laboratory discussion.

Answers to Questions

Percent change in mass

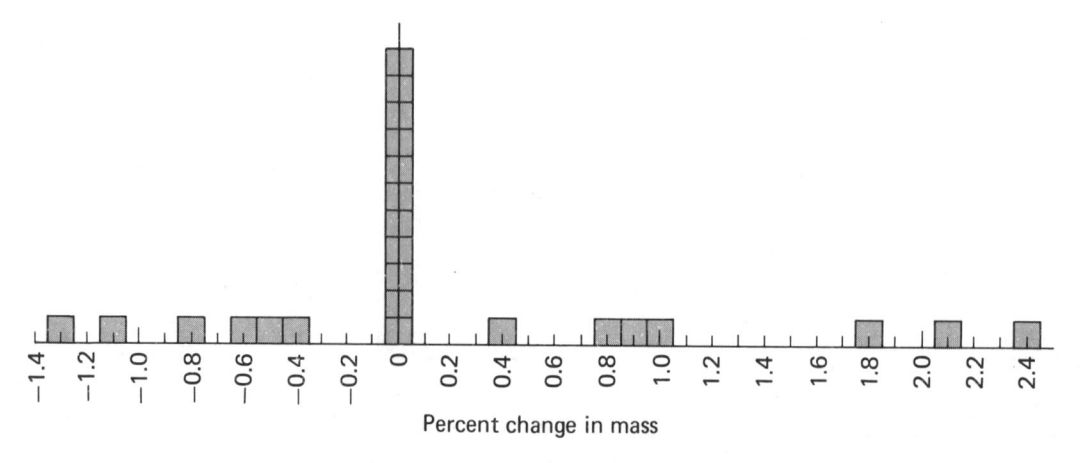

Figure VIII

Again, as Fig. VIII shows, the result most frequently obtained is zero. This time the results are nearly evenly distributed on either side of the zero (the value −10.3 percent in the following table is clearly in error and was not histogrammed).

Mass of Ice (g)	Mass of Water (g)	Change in Mass (g)	Percent Change
5.40	5.38	−0.02	−0.4
6.83	6.83	0.00	0.0
9.10	9.05	−0.05	−0.5
6.64	6.55	−0.09	−1.3
7.14	7.10	−0.04	−0.6
6.43	6.48	+0.05	+0.8
14.95	14.95	0.00	0.0
7.24	7.24	0.00	0.0
11.14	11.14	0.00	0.0
7.40	7.40	0.00	0.0
6.83	6.83	0.00	0.0
4.45	4.45	0.00	0.0
3.64	3.64	0.00	0.0
1.26	1.29	+0.03	+2.4
2.45	2.43	−0.02	−0.8
2.52	2.26	−0.26	−10.3
3.29	3.29	0.00	0.0
8.21	8.12	−0.09	−1.1
4.69	4.71	+0.02	+0.4
2.90	2.93	+0.03	+1.0
10.48	10.67	+0.19	+1.8
3.31	3.38	+0.07	+2.1
3.26	3.26	0.00	0.0
6.50	6.50	0.00	0.0

Apparatus and Materials

Balance
Small plastic bottle with cover

Ice cube
Paper towel

EXPERIMENT: THE MASS OF MIXED SOLUTIONS 2.11

Your students discover in this experiment that within the experimental accuracy of the balance, mass does not change when a solid is formed by mixing two liquids. A new substance is formed when the solutions are mixed, but this is not the important point. It is sufficient for students to recognize only that the change that takes place is different from the changes in the two previous experiments.

The Experiment

All students should be provided with lead nitrate solution. Enough for 25 experiments can be made by dissolving 15 g of solid in 100 cm^3 of water.

The second solution can be either sodium iodide, sodium chloride, copper sulfate, or Epsom salt solution. Only one of these is essential, but for dramatic effect you can give a quarter of your students sodium iodide solution, another quarter sodium chloride solution, another quarter copper sulfate solution, and the remainder Epsom salt solution.

All solutions can be made up with the same concentration as the lead nitrate solution. After preparing the solutions mix a few cubic centimeters of lead nitrate solution with each of the remaining solutions to be sure that the solutions have been made correctly and that a precipitate forms.

The precipitates are respectively lead iodide, lead chloride, and lead sulfate. Students may dispose of the solid by flushing it down the sink drain with plenty of running water.

Answers to Questions

Figure IX represents a typical set of classroom data. All but a few of the results are either zero or within 0.02 g of zero.

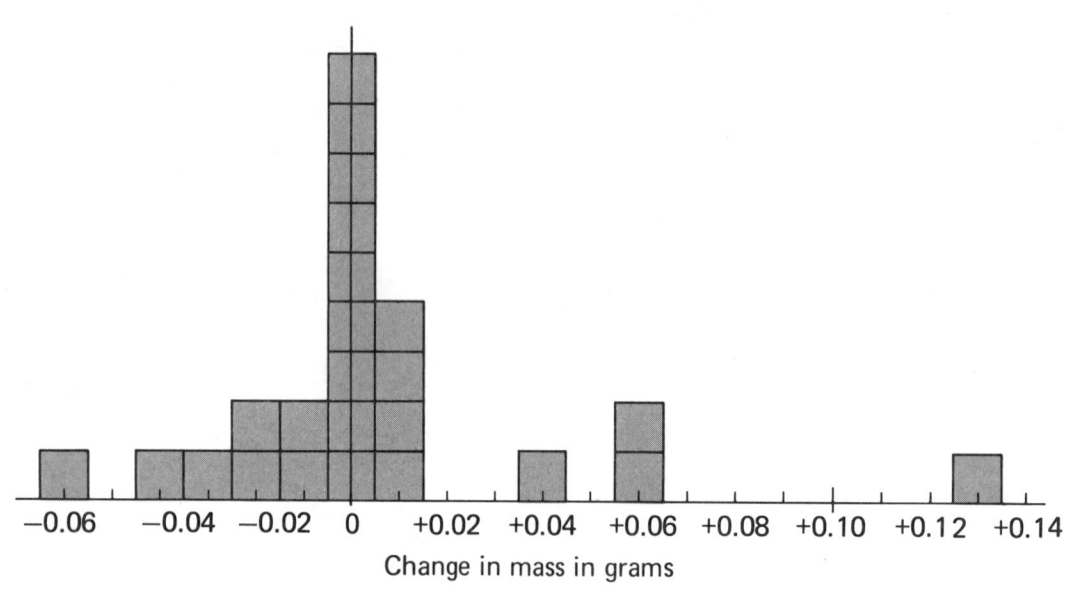

Figure IX

Apparatus and Materials

Balance
2 Plastic bottles with covers
Lead nitrate solution
Paper towels
Test-tube brush

At least one of the following:
Sodium iodide solution
Sodium chloride solution
Copper sulfate solution
Epsom salt solution

Hot, soapy water and a test-tube brush are necessary in removing the solid that remains in the bottles at the end of the experiment.

EXPERIMENT: THE MASS OF COPPER AND SULFUR   2.12

The purpose of this experiment is to show that mass is conserved even when a rather violent change takes place. Again, we are not concerned with the chemical change as such. We are concerned only with the effect the change has on the mass. In fact, it is unwise to make any distinction between chemical and other changes. Typical results are shown in Fig. X.

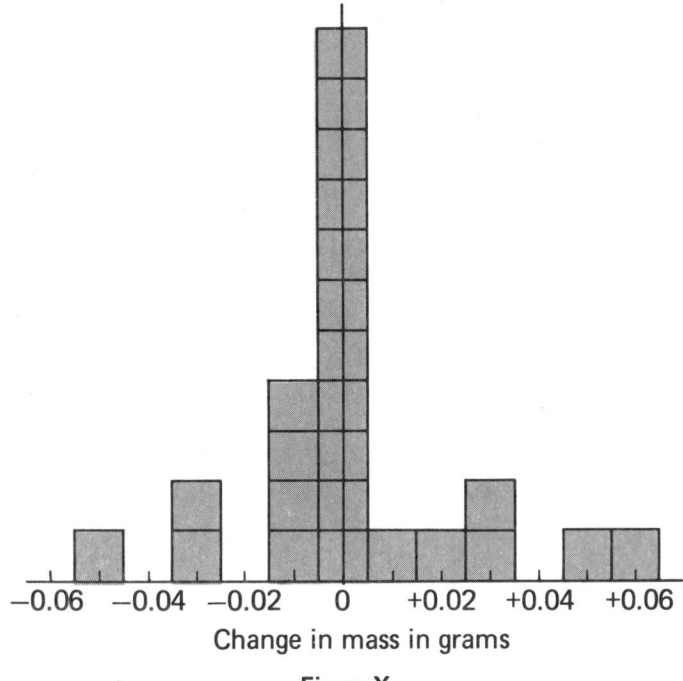

Figure X

Your students will not know if the mass of copper and sulfur will change when these substances are heated together, but they probably will expect from the results of the last three experiments that no change in mass will be observed.

The Experiment

The granular copper and sulfur should be massed and placed in a test tube. (Do *not* use copper dust or powder, which reacts too rapidly.) The two components must be mixed thoroughly before heating if the reaction is to be visible. To prevent the loss of any material, the mouth of the test tube is covered with a piece of rubber sheet and is tightly sealed with a rubber band. No gas is produced in this experiment. The rubber sheet allows the air in the tube to expand while it is hot. The sheet should be flat when the tube has cooled back to room temperature.

When test tubes are being heated, their mouths should always be pointed in a direction away from students. A vigorous reaction should take place after about five minutes of heating. After the reaction begins, the burner should be removed and the test tube left in position on the pegboard for several minutes until it is cool enough to handle and mass. Students must wear their safety goggles during this experiment.

Answers to Questions

The mass of the copper and sulfur does not change by an amount greater than that of the precision of the balance.

There appears to be a new substance formed that is not like either copper or sulfur. (Do not let your students be concerned with the name or composition of the new substance, only that there has been a change.)

Apparatus and Materials

Test tube	Copper (granulated, fine), 2 g
Rubber sheet	Towel
Rubber band	Matches
Alcohol burner	Rubbish container in which to
Pegboard and clamps	discard test tubes and contents
Balance	One hammer per class
Sulfur (powder), 1 g	Safety glasses

Because the solid remaining in the test tube at the end of the experiment is difficult to remove, the test tube may have to be discarded. In this case, have your students break it to examine the solid. As a precaution against flying glass, the test tube should be wrapped in a paper towel and tapped with a hammer.

You may be able to get the fused solid slug free from the tube after it has cooled by moving the burner along the tube from the closed end about halfway to the stopper. This drives away the excess sulfur, which very often is what cements the slug in place.

2.13 EXPERIMENT: THE MASS OF GAS

There are two important things to be learned in this experiment. First, mass does not change in a reaction that produces a gas, provided no gas is allowed to escape; this fact further builds up the evidence for conservation of mass. Second, conservation of mass can account for the loss in mass when gas is allowed to escape from the capped bottle in which it is generated, thus showing that a gas has mass—a fact not appreciated for many years.

The Experiment

To get good results it is necessary (1) to have bottles with caps that will not pop off or leak gas when the pressure builds up inside, and (2) to work quickly in closing the cap after the small piece of Alka-Seltzer tablet is dropped into the water. The size of the bottle used is quite critical. The bottle should be dry before massing. The cap should be loosened slowly so that no water bubbles out when the gas escapes. The change in mass when the gas escapes is small, about 0.06 g, but great enough to be measured on the balance.

This technique for measuring the mass of gas produced in a reaction by using the conservation of mass will be used again in Chapter 3 when students measure the density of a gas.

To avoid any chance that students will use more than ⅛ tablet, divide and dispense the Alka-Seltzer yourself. The ⅛ tablet is *mandatory* to prevent too great a pressure build-up and consequent explosion of the bottle. USE ONLY THE BOTTLE SUPPLIED or one you know will not break under pressure. As an added safety precaution, it is suggested that the bottle be wrapped with tape. *Safety glasses must be worn.*

Sample Data

Initial mass (bottle, cap, water, and ⅛ tablet)	51.48 g
Final mass (bottle, cap, and solution)	51.48 g
Mass after gas is released	51.40 g

Answers to Questions

By this time you can expect students to say that if a solid and a liquid produce a gas, there will be no mass change, but to show this you must contain the gas in a closed system and not let it escape. However, some students may not realize that a gas has mass and therefore not consider it important to prevent the gas from escaping.

When the cap is tightly on the bottle, the production of gas inside does not affect the mass of the bottle and contents.

One can hear the gas escaping from the bottle when the cap is loosened slowly.

When the gas is allowed to escape, there is also a noticeable loss in mass. With ⅛ Alka-Seltzer tablet the loss is about 0.08 g.

Apparatus and Materials

Balance
Pegboard and clamps
Small, thick-walled bottle with
    tight-fitting cap

⅛ Alka-Seltzer tablet
Paper towel
Safety glasses
Tape

THE CONSERVATION OF MASS   2.14

LAWS OF NATURE   2.15

These sections summarize the whole chapter and lay the groundwork for the understanding of other laws of nature to be studied later. Reading them aloud in class and discussing them with your students as you go along is strongly recommended.

Powers of 10 are mentioned for the first time in the footnote to Sec. 2.14. This notation is used at the beginning as a shorthand for positive powers only. In the next chapter negative powers are introduced, but students can do all their calculations in regular notation. Only in Chap. 10 will it be necessary to make extensive calculations with powers of 10. Develop skills in using this notation gradually. Using powers of 10 in your own calculations on the chalkboard will help students to get used to it.

CHAPTER 2—ANSWERS TO PROBLEMS

The table below classifies problems according to their estimated level of difficulty and the sections to which they relate. The column headed "Class Discussion" lists questions that appear in the first three columns—some quantitative, others not—which can either be brought up in class directly or first be assigned and then be discussed. In addition to the questions listed in the last column (Home or Lab), there may be others that you will want to extend into lab or home experiments. The answers to questions marked with a dagger (†) are given in the appendix of the text as well as in this Teacher's Guide.

Section	Easy	Medium	Hard	Class Discussion	Home or Lab
1	1†	3, 26	2, 27		
2	5†	4, 6, 7, 8		6	
3	9†, 29	10, 11, 30		28, 30	
4-5	12, 33, 37	13, 32, 34, 41	35, 45	32, 33, 45	31, 41
6-8	14†, 36, 40	15, 18	16, 17, 38	16, 39, 40, 44	39, 44
9	19†, 20†				
10-12	22	21†, 42		42	43
13-14	23	24			
15		25			

1† A student has a large number of cubes that measure 1 cm along an edge. If you find it helpful, use a drawing or a set of cubes to answer the following questions.
   a) How many cubes will be needed to build a cube that is 2 cm along an edge?
   b) How many cubes will be needed to build a cube that is 3 cm along an edge?
   c) Express, in cubic centimeters, the volumes of the cubes built in (a) and (b).

   a) <u>8 cubes.</u>
   b) <u>27 cubes.</u>
   c) <u>8cm^3; 27 cm^3.</u>

2 A student can carry two filled grocery bags. What is the approximate volume of each of these bags? How many cereal boxes can fit into each bag? What assumptions do you have to make to fit the boxes in the bag?

Your students will probably find this question difficult, for they are not likely to be accustomed to making estimates and assumptions. Both grocery bags and cereal boxes come in a wide variety of sizes and shapes, although almost all the shapes are rectangular. The student will have to make arbitrary selections of bag size and box size; some of the students will no doubt make some measurements on bags and boxes at home, but all students are familiar enough with them so that any reasonable estimate of their sizes will be suitable here.

The students will have to assume that all the cereal boxes are the same size and that they have found the way to arrange the boxes in the bag so as to fill the space as completely as possible.

The bottom of a representative grocery bag might measure 30 cm in length by 17 cm in width, and its height might be 43 cm. The nearer the dimensions of the cereal boxes are to being integral submultiples (½, ⅓ ... etc.) of the dimensions of the bag, the closer will be the fit. For example, if Boppies are sold in packages 8 cm by 15 cm by 21 cm, eight such packages will nearly fill the bag. The student could state that the volume of the bag was a bit more than 8 Boppie boxes.

If Krax come in boxes 16 cm by 4 cm by 20 cm, the students could fit 14 boxes into the bag; they might estimate that if they could have half-boxes or third-boxes, 2 or more might be able to fit into the bag. The bag could then be described as having a volume of 16 to 19 Krax boxes.

3 Figure A shows a graduate, used for measuring the volume of liquids, that is shaped like a cone. Why are the divisions not equally spaced?

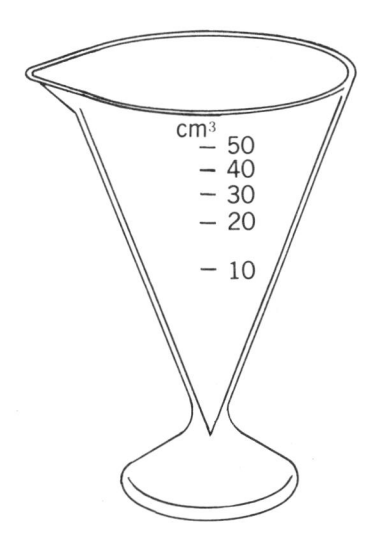

Figure A

One can think of each cm³ of liquid added to the graduate as forming a disk, whose volume is equal to the product of its depth and its surface area. As the level of the liquid rises, the area of its surface increases and consequently the depth of each cm³ layer of liquid grows smaller. Hence the divisions indicating equal volumes crowd more closely together as the quantity of liquid increases.

28 Volume and Mass

4 The scale in Fig. B is in centimeters. Estimate the positions of arrows *a* and *b* to the nearest 0.1 cm. Can you estimate their positions to 0.01 cm? To 0.001 cm?

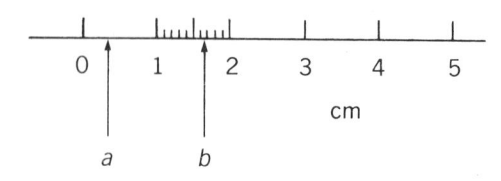

Figure B

The estimate of the position of arrow *a* is 0.3 cm or 0.4 cm and of arrow *b* is 1.66 cm or 1.67 cm. It is impossible to estimate the position of arrow *a* to the nearest 0.01 cm. The position of arrow *b* can be estimated to about the nearest 0.02 cm. The position of neither *a* nor *b* can be estimated to the nearest 0.001 cm.

This problem is intended to show students that it is possible to read a scale graduated in millimeters to about 0.02 cm. Beginning students are prone to read a centimeter scale only to the nearest division.

5† What fraction of a cubic centimeter do the smallest divisions on each of the graduated cylinders in Fig. C represent?

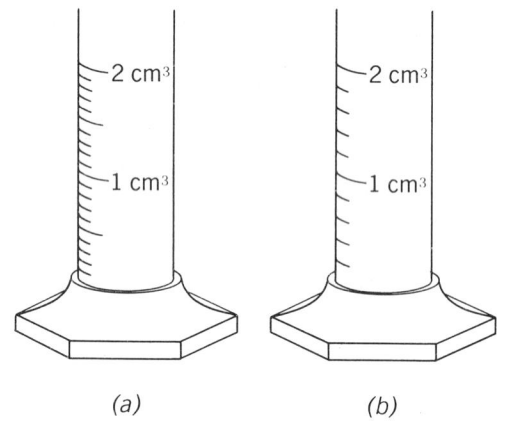

Figure C

a) 0.1 cm^3.
b) 0.2 cm^3.

6 A close look at Fig. D shows that the top of the liquid contained in the graduated cylinder is not flat but curved. How do you decide how much water is in the cylinder?

This question should be assigned as preparation for Expt. 2.3. A class discussion is necessary, since it will be difficult for students to decide for themselves how to read a graduated cylinder. Even if they cannot decide on the correct answer, it is worthwhile to have them worry about the problem and realize its importance before the answer is

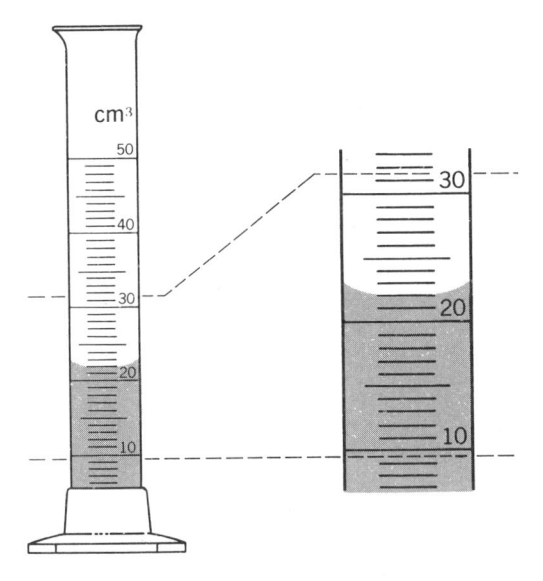

Figure D

given to them. Generally we agree to read the volume of the liquid by looking horizontally at the surface and reading the position of its lowest part. This introduces a smaller error than does reading the top of the meniscus, and the error becomes less as the diameter of the measuring cylinder is increased. Students should form the habit of always reading the position of the bottom of the meniscus. However, if we are interested only in differences in volume (as is the case in measuring the volume of a solid by water displacement, for example), it makes no difference how the volume is read as long as it is read in the same way both times.

With liquids that do not wet glass, the meniscus curves down at the edges, and the top of the meniscus should be read. Mercury is the common example of such a liquid, but students will not measure mercury in a graduated cylinder in this course. In any case, we read the level of the liquid as it is at the center of the cylinder, away from the walls.

7 **Three students reported the length of a pencil to be 12 cm, 12.0 cm, and 12.00 cm. Do all three readings contain the same information?**

No. A measurement reported to be 12 cm tells us that the pencil's length is closer to 12 cm that it is to 11 cm or 13 cm; that is, its length could be anything between 11.5 and 12.5 cm.

Similarly, when the length is reported to be 12.0 cm, the measurement tells us that the pencil's length is closer to 12.0 cm that it is to 11.9 cm or 12.1 cm; the length could be anything between 11.95 cm and 12.05 cm.

Finally, the length reported as 12.00 cm is closer to 12.00 cm than it is to 11.99 cm or 12.01 cm,; it can be anything between 11.995 cm and 12.005 cm. This length reported as 12.00 cm has the smallest uncertainty of the three.

30 Volume and Mass

8 What advantage is there to making graduated cylinders narrow and tall rather than short and wide?

Each cm³ of liquid added to the graduate forms a disk whose volume is the product of its surface area and its depth. In the narrow cylinder the surface area will be smaller and hence the depth will be greater. The dimensions indicating cm³ will be farther apart; they will be easier to read, and a more accurate estimation can be made between them.

9† A volume of 50 cm³ of dry sand is added to 30 cm³ of water for a total volume of 60 cm³.
a) What is the volume of water that does *not* go into air spaces between the sand particles?
b) What is the volume of water that does fill air spaces between the sand particles?
c) What is the volume of the air spaces between the particles in dry sand?
d) What is the volume of the sand particles alone?
e) What fraction of the total volume of the dry sand is sand particles?

a) 10 cm³.
b) 20 cm³.
c) 20 cm³.
d) 30 cm³.
e) 0.60.

10 How would you measure the volume of granulated sugar?

Methods similar to those used in Expt. 2.3 for measuring the volume of sand can be used, but with a liquid in which sugar does not dissolve. Sugar has a very low solubility in grain alcohol, but students will not learn this until they do experiments in Chap. 4 on the solubility in alcohol of various substances. Sugar is insoluble in ether or chloroform, but these liquids are too dangerous for student use.

11 How would you measure the volume of a cork stopper?

You could measure the volume by the displacement of water; push the cork below the water surface with a long pin. As an alternative you could tie the cork to a stone by a short length of thread, put water in a graduated cylinder, take a reading with just the stone submerged, and then a reading with both under water. The difference in reading is the volume of the cork stopper.

12 When you buy things at the store, are they measured more often by volume or by mass? Give some examples.

In general, liquids are sold by volume (*e.g.*, a gallon of milk, a pint of vinegar) and solids by mass (*e.g.*, 10 pounds of sugar, 5 pounds of flour). However, the contents of canned goods are frequently expressed in both ways, and in most places it is required that the net mass of the contents be marked on the container, which indicates that mass is a better measure of amount than volume.

13   What is your mass in kg?

Divide your mass by $2.2 \frac{lbs}{kg}$. If you weigh 110 lbs, your mass is $\frac{110 \text{ lbs}}{2.2 \text{ lbs/kg}} = 50$ kg.

14 † Karen Jones massed an object three different times, using the same balance and gram masses. She obtained the following values: 18.324 g, 18.308 g, and 18.342 g. How could she best report the mass of the object?

Karen can best report the mass of the object as the average of her three values:

$$\begin{array}{r} 18.324 \text{ g} \\ 18.308 \text{ g} \\ \underline{18.342 \text{ g}} \\ 3 \overline{)54.974} \\ \hline 18.325 \text{ g} \end{array}$$

This is as far as a student would be expected to go. One could find the average deviation of the measurements:

$$\begin{array}{r} 18.325 - 18.324 = 0.001 \text{ g} \\ 18.325 - 18.308 = 0.017 \text{ g} \\ 18.342 - 18.325 = \underline{0.017 \text{ g}} \\ 3 \overline{)0.035 \text{ g}} \\ \hline 0.012 \text{ g} \end{array}$$

and express the mass as 18.325 ± 0.012 g. This manner of writing the mass conveys more information than simply stating the average. If you have any students interested in mathematics, they might like to know about this.

15   A student masses an object on his balance. By mistake he places the object in the pan on the same side as the rider. He balances the object by means of 4.500 g in the opposite pan and by setting the rider to 0.060 g. What is the mass of the object?

The mass represented by the rider must be subtracted from the 4.500 g. The mass of the object is 4.500 g − 0.060 g = 4.440 g.

16 Five students in turn used the same balance to measure the mass of a small dish; none knew what results the others obtained. The masses they found were

Student	Mass (g)
1	3.752
2	3.755
3	3.752
4	3.756
5	3.760

a) Can you tell whether any student made an incorrect measurement?
b) Do you think there is anything wrong with the balance?
c) What do you think would be the best way to report the mass of the dish?
d) How precise do you think the measurements were?

a) The only basis for judgment lies in the student's observations in Expt. 2.8, The Precision of the Balance. There they saw variations similar to those observed here, although these variations may be somewhat larger. However, they should see no basis for thinking that there was an incorrect measurement.

b) The students again have no other basis for judging except their observations in Expt. 2.8. There appears to be no reason for thinking anything was wrong with the balance.

c) The best way to report the mass of the dish would be to give the average of the readings:

$$\begin{array}{r} 3.752 \\ 3.755 \\ 3.752 \\ 3.756 \\ 3.760 \\ \hline 5 \overline{\smash{)}18.775} \\ 3.755 \text{ g} \end{array}$$

d) Students may not know exactly what is meant here; some may say "to better than 0.01 g," which is a satisfactory answer. An interested student might follow the suggestion in the discussion of Prob. 14 and find the average deviation of the measurements from the average mass; this is 0.002 g. The student could then express the mass of the dish as 3.755 ± 0.002 g. The uncertainty in the average mass is less than 0.1 percent.

17 An equal-arm balance good to 0.01 g is used to mass two objects. If their masses are measured as 0.10 g and 4.00 g, what is the expected percentage error of each measurement?

An error of 0.01 g in a measurement of 0.1 g is 0.01/0.1 = 0.10 = <u>10 percent.</u>
An error of 0.01 g in a measurement of 4.00 g is 0.01/4.00 = 0.0025 = <u>0.25 percent.</u>

18 You were told that your balance is an equal-arm balance, but suppose that the right arm is longer than the left arm. Would the object that is massed on your balance appear to have a mass that is greater than, less than, or the same as its true mass?

The object to be measured is placed on the left pan and the known masses on the right. The right arm would require less mass to balance the object when the right arm is longer than the left arm, and the object would appear to have less mass than its true mass.
  You can point out to the students that they have probably noticed this effect when they played on a seesaw with a friend. However, it is easy enough to demonstrate. Suspend unequal masses from a straight stick—say, a meter stick—and support it with your finger. By sliding it back and forth, you can find the point where it balances. You will see the smaller mass suspended from the larger segment of the stick.

19† In Expt. 2.9, how could you recover the dissolved salt? How do you think its mass would compare with the mass of dry salt you started with?

Boil away or evaporate the water. The mass of the salt recovered would be the same as at the beginning.

20† If the change in mass of the salt and water solutions was −0.0001 g in Expt. 2.9, would you have observed this change using your balance?

No. This change represents a change of the position of the rider of 1/100 of the distance between the calibration marks on the balance arm. You could not observe a change as small as this on your balance.

34 Volume and Mass

21† The following data were obtained in an experiment in which copper and sulfur were reacted.

	Mass (g)
Tube and cover	20.484
Tube, cover, copper, and sulfur before reaction	23.440
Tube, cover, copper, and products after reaction	23.386

a) What is the mass of the substances before the reaction?
b) What is the apparent change in mass of the reacting substances?
c) What is the apparent percentage change in mass of the reacting substances?

a) 2.956 g
b) 0.054 g
c) About 2 percent (1.8 percent)

22 A test tube having 4.00 g of iron and 2.40 g of sulfur was heated in a manner similar to that of the copper-and-sulfur experiment. The total mass of the tube and contents measured on the balance before the heating was 36.50 g. After the heating, its mass was measured again. The mass of the tube and contents was 36.48 g.
a) Are you inclined to think it reasonable that mass was conserved during this experiment?
b) What additional steps would you take to strengthen your inclination?

a) The mass of the test tube and its contents before and after the experiment was nearly the same. If different gram masses were used in the two massings, we would be inclined to think that mass had been conserved.
b) Our inclination would be strengthened if similar results were obtained as a result of a number of repetitions of the experiment, preferably with varying amounts of iron and sulfur. It would also be strengthened if we knew the balance to be capable of massing to a smaller fraction of a gram, and if the comparison of the mass of the test tube and its contents before and after heating showed differences no greater than those consistent with the accuracy of the balance.

23 a) Express the following numbers in powers of 10.
    100   10,000   100,000,000
b) Write the following numbers without using exponents.
    10^5 10^6 10^9

a) 10^2 10^4 10^8
b) 100,000   1,000,000   1,000,000,000

24 a) Express the following numbers in powers of 10.
1,000   5,280   93,000   690,000
b) Write the following numbers without using powers of 10.
5.0 × 10³   10⁷   1.07 × 10²   4.95 × 10⁴

a) 10³   5.28 × 10³   9.3 × 10⁴   6.9 × 10⁵
b) 5,000   10,000,000   107   49,500

25 There is an old saying, "Whatever goes up must come down." Does this express a law of nature? Why, or why not?

This is an open-ended question that should provoke considerable discussion. In the text a law of nature has been defined as "... a guessed generalization based on experiments, often crude experiments." This would qualify the saying as a law of nature, for it is indeed based on countless experiments since people first began to describe their environment through observation and generalization. The experiments are crude, of course, and so is the statement of the law; but this does not disqualify it as a law. Classroom discussion will probably suggest ways in which the statement could be improved.
Only one confirmed and repeated contradiction is needed to necessitate a change in a law of nature; but no finite number of experiments can ever prove the law to be true for all possible conditions. Indeed, some students may point out a case in which this law does not apply—namely, the escape of spacecraft from the earth's gravitational field. They may wish to modify the statement of the law to show for what conditions it holds.
No matter how carefully a natural law is stated, no matter how competent the scientist is who announces it, there never is any guarantee that it will not be later modified to conform to new observations or new interpretations of old observations; but it will still apply in the realm in which it has been tested and found correct.

26 How would the volume of a piece of glass as measured by displacement of water compare with its volume as measured by displacement of burner fuel?

The volume would be the same measured by either method. The glass is not soluble in either liquid, the space taken up by the glass is the same in both water and burner fuel, and the glass has to push the same volume of liquid out of the way in either case.

27 In determining the volume of a rectangular box, five cubes were found to fit exactly along one edge, and four cubes fit exactly along another edge. However, after six horizontal layers had been stacked in the box, a space at the top was left unfilled.
a) If the height of the space was half the length of a unit cube, what was the volume of the box?

36 Volume and Mass

b) If the height of the space was 0.23 of the length of a unit cube, what was the volume of the box?

a) The diagram in Fig. XI shows that each cube in the top layer of the unit cubes that fit in the box should have a half cube on top of it in order to fill the box; therefore the volume of the box is (5 × 4 × 6) + (5 × 4 × ½) = 120 + 10 = 130 cubes.

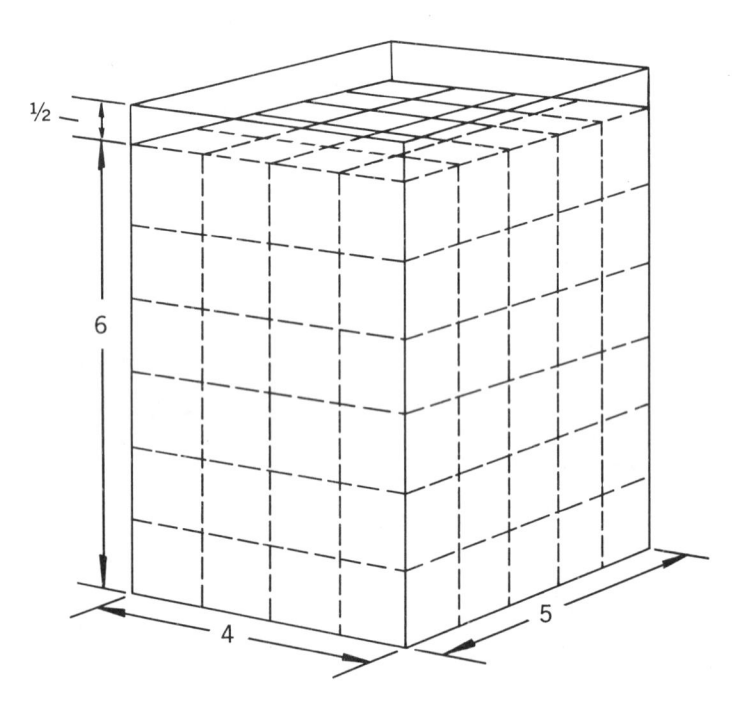

Figure XI

b) In like manner, the volume of the box would be (5 × 4 × 6) + (5 × 4 × 0.23) = 124.6 cubes. It can now be pointed out that since the height of the box is 6.23 times the side of the cube, the volume can be more simply calculated as 5 × 4 × 6.23 = 124.6 cubes. This problem is designed to help students understand that when unit cubes do not exactly fit into a given rectangular volume, the number (and fraction) that does fit can be calculated by taking the product of the three linear dimensions—something that is not discussed in the text.

28 What is the total number of cubes that will fit in the space enclosed by the dashed lines in Fig. E? Is there more than one way to find an answer?

The dimensions of the space enclosed by the dashed lines is 2 cubes × 3 cubes × 4 cubes. Multiplying these dimensions gives 24 unit cubes. Another way is to count the total number of cubes needed to fill the rear side of the space and then doubling this, because the space is 2 cubes thick. You can see that there are four cubes along the bottom, and it will take three layers to fill the back side, or a total of 12. Doubling this gives 24.

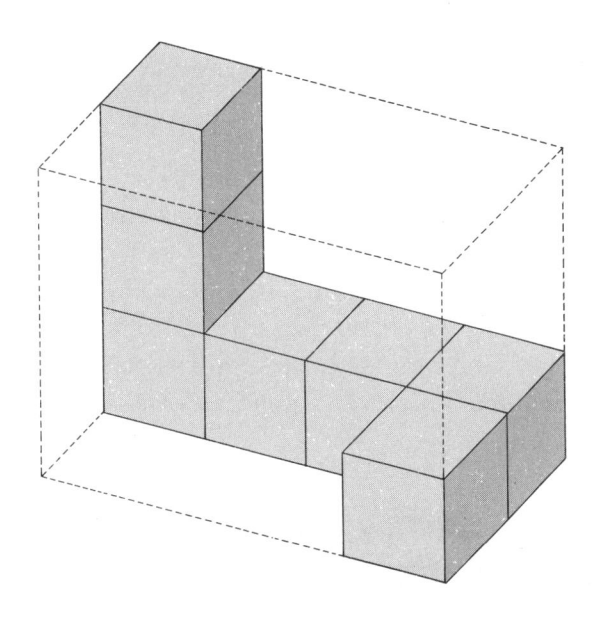

Figure E

29 In an experiment in which the volume of dry sand is measured by the displacement of water, the sand was slightly wet to begin with. What effect would this have on the volume of air space that was calculated? On the percentage of the volume that was air space?

Wet sand already has part of the space between the particles filled with water. This means it will take less water to fill the air space, and consequently the calculated volume of air space and the percentage of the volume that was air space would both be less than in the case of dry sand.

30 a) How would you measure the volume of a sponge?
    b) What have you actually measured by your method?
    c) Does this differ from your measurement of the volume of sand?

a) Case I: If the sponge is rectangular, you could measure the dimensions and calculate the volume.
    Case II: You could use water displacement. If the sponge is too large to fit in a graduated cylinder, put water in any container large enough, and mark the water level. Submerge the sponge and squeeze the air out of it. Then mark the water level again. On removing the sponge, you would find how much water was needed to make up the difference in levels by filling the container from one level to the other with water from a graduated cylinder.
b) Case I: You have measured the space taken up by the sponge as a whole.

Case II: You have measured the volume only of the sponge material that is impervious to water.
c) The method in Case I gives the same kind of result as that used in the measurement of dry sand. The procedure in Case II is in principle the same as that used in determining the volume of wet sand.

31 a) Completely fill two small bottles with water. Pour the water into a single larger vessel. Now refill the bottles with the same water. Are they both filled completely?
b) Now do the same thing again, but fill one bottle with water and the other with burner fuel. Compare the total volume of burner fuel and water before and after they were mixed together and poured back into the bottles. Is volume a good measure for the quantity of matter in this case?

These questions can be answered only by doing an experiment, either in the laboratory or at home. Merely guessing at the answers is valueless.

a) As nearly as one can tell, both bottles are still full of water. Of course, a few drops have adhered to the single larger vessel, but the loss in volume is so small as to be scarcely noticeable.
b) This time the mixture of the two liquids definitely will not quite fill the two bottles. The volume of a mixture of water and alcohol (methanol was used) is about 4 percent less than the sum of the volumes of the water and alcohol separately. Volume is not a good measure for the quantity of matter in this case.

32 Fuel oil usually is sold by the gallon, gas for cooking by the cubic foot, and coal by the ton. What are the advantages of selling the first two by volume and the last by mass?

Fuel oil and fuel gas are conveniently measured in terms of volume units, and the fuel value per dollar will be the same for all such units of each fuel if pressure and temperature effects are not excessive. The volume of oil is nearly independent of pressure, and it changes by only about 0.3 percent over a 30°C temperature range. Gases are much more affected by changes in pressure and temperature. However, fluctuations in gas pressure used for cooking are usually too small to cause significant effects. If the gas meter operates where the average temperature is much different from that for which the meter is designed (usually 60°F), a correction is made.

Coal is not so easily measured in volume units. Scales large enough to mass coal trucks and railroad cars directly are commonly used; with the mass of the empty truck known, one can get an accurate measure of the mass of the coal.

33 In the following list of ingredients for a recipe, which are measured by volume, which by mass, and which by other means?

1½ pounds ground chuck   pinch of pepper
1 medium-size onion   3 drops Worcestershire Sauce
½ cup chopped green pepper   oregano to taste
4 slices day-old bread   3 tablespoons oil
1 teaspoon salt   1 1-pound can tomato sauce

Obviously measured by mass are 1½ pounds of chuck and 1 pound of tomato sauce. Obviously measured by volume are ½ cup of chopped green pepper, 1 teaspoon of salt, 3 drops of Worcestershire sauce, and 3 tablespoons of oil.

The bread and onion are measured by a simple count; one assumes that the mass of an onion and the mass of a slice of bread will not vary enough to matter in this case. The pinch of pepper is a volume unit. The unit for oregano does not correspond to any of the ways of measuring matter that we have studied!

34 a) What is the volume of an aluminum cube whose sides are 10 cm long?
    b) What is the mass of the aluminum cube? (One cubic centimeter of aluminum has a mass of 2.7 g.)

a) Volume = length × width × height = 10 cm × 10 cm × 10 cm = 1,000 cm³.
b) If 1 cm³ of aluminum has a mass of 2.7 g, then 1,000 cm³ will have a mass of 1,000 times as much: 2.7 g × 1,000 = 2,700 g.

35 One cubic centimeter of gold has a mass of 19 g.
    a) What is the mass of a gold bar 1.0 cm × 2.0 cm × 25 cm?
    b) How many of these bars could you carry?

a) The volume of the bar = 1.0 cm × 2.0 cm × 25 cm = 50 cm³; if 1 cm³ of gold has a mass of 19 g, then 50 cm³ will have a mass 50 times as much or 19 g × 50 = 950 g = 0.95 kg.
b) The mass of a single bar is approximately 1 kg, which can also be expressed as about 2 pounds. How many of these bars can be carried depends on how strong the student thinks he or she is and the distance they are to be carried. Perhaps 40 to 100 bars is a good answer.

36 Suppose that you took a balance home. When you were ready to use it, you found that you had forgotten a set of gram masses.
    a) How could you make a set of uniform masses from materials likely to be found in your home?
    b) How could you relate your unit of mass to a gram?

40 Volume and Mass

a) You would locate a set of similar objects (paper clips, small nails, etc.) and test them with your balance for uniformity of mass. Then you could measure the masses of various objects in terms of the numbers of your new standard masses required to balance them.

b) To get the relation between one gram and one of your personal standard masses, you would have to compare your new set with the gram masses on a balance.

37 Figure 10.10 (p.196) shows a sensitive equal-arm balance. Suggest reasons why it is enclosed in a case and why it is used with the sliding front cover closed.

The case will prevent dust and dirt from settling on the balance. Keeping the cover closed will prevent air currents from affecting the swing of the balance.

38 You wanted to find the mass of water in a plastic bottle, and you took the following measurements using your equal-arm balance.

Mass of bottle and water	21.48 g
Mass of empty bottle	9.56 g
Mass of water	11.92 g

After you completed your measurements and calculations, you saw that you forgot to set the left-hand rider correctly; the beam was not level when the right-hand rider was on the zero mark and nothing was on the pans of the balance. Must you repeat the measurements to obtain the mass of the water?

No. The mass of the water is obtained by subtracting the mass of the empty bottle from the mass of the bottle and water. The error made by not setting the left-hand rider correctly changed both these measurements by the same amount, and the error was subtracted out when you subtracted the mass of the empty bottle from the mass of the bottle and water. However, the only correct mass is the mass of the water.

39 a) You can compare your standard masses on the equal-arm balance after you have carefully adjusted it. How closely is the 50-gram mass equal to the sum of the 20-, 10-, 10-, 5-, 2-, 2-, and 1-gram masses? How closely is the 5-gram mass equal to the sum of the two 2-gram masses and the 1-gram mass? You can make other comparisons as well.

b) When you are finding the difference in mass between an empty container and the container filled with liquid, why should you try to use as nearly as possible the same particular masses for both measurements?

a) Depending on the particular mass set you have, you may find no measurable difference between the mass of the 50-g mass and the sum of the others given, or you may find that there is a difference that might be as much as 0.09 g. This is only about 0.2 percent. The mass of the 5-g mass may differ from the sum of the others given by as much as 0.003 g, less than 0.1 percent.

b) Because there may be small variations in the masses of the mass set, you should try to use as nearly as possible the same particular masses so that you do not alter the difference in mass you are measuring by accidental differences caused by the standard masses themselves.

40 Suppose you lost the rider for your scale. Try to think of another method, not using a rider, by which you could measure hundredths of a gram.

Probably the simplest thing to do is to trim a piece of fine wire or string until it balances 0.1 g, then divide it into ten equal parts. Or a narrow strip of thin paper could be trimmed until it balances 0.1 g, and then it could be divided into ten equal parts. Graph paper is suitable for this, as the lines will assist in the dividing process.

41 Suppose you balance a piece of modeling clay on the balance. Then you reshape it. Will it still balance? If you shape it into a hollow sphere, will it still balance?

The piece of modeling clay will still balance (as long as none of the clay has stuck to the fingers!), since the mass has not changed.

If the clay is shaped into a hollow ball, it will still balance, since the mass has not changed. Some students may worry about the air trapped inside the sphere; this is exactly offset by the increased buoyant effect of the displaced air, so that it does not matter. However, do not get distracted into a discussion of Archimedes' principle. Have doubtful students try this in the laboratory.

42 Suggest a reason for putting the lid on the small container that you used in studying the mass of ice and water.

If the container remained cold throughout the experiment, the lid would prevent condensation of atmospheric moisture from taking place in the inside of the container, where it could not be wiped off.

43 In Expt. 2.10 would the mass of the container and its contents stay the same if you started with water and froze it? Try it.

A student would probably guess the correct answer: that the mass would remain constant. When one tries the experiment, some precautions may be necessary, Moisture may condense and freeze on the outside of the can when it is cooled in a freezer not equipped with an automatic defroster.

44 **Estimate in grams the mass of a wristwatch. Now find the mass of a nickel (5¢) on your balance. Estimate the mass of the watch again. Did you change your estimate? Does knowing the mass of a nickel help you to better estimate your own mass? Why?**

By now your students will have some experience with gram masses in the laboratory, but they are unlikely to be able to estimate closely the mass of an object such as a wristwatch. Usually the tendency is to overestimate the mass. Some students may think they have no idea at all of its mass; but by suggesting ridiculous masses—such as 2000 g or 0.1 g—you can get them to decide on some value. Here one should be satisfied with an order-of-magnitude figure—that is, one that falls somewhere between 1/10 and 10 times the true mass.

After the students mass the nickel—its mass is somewhere in the neighborhood of 5 g—their revised estimates will be generally closer to the true mass of the watch. They then have a way of comparing the mass of the watch with an object of known mass.

They are less likely to be helped in the estimate of their own mass by knowing the mass of a nickel; the masses of a person and the mass of a nickel are too far apart. In any case, they are likely to know their mass in metric units already, as a result of answering problem 13.

(Incidentally, it is interesting to note that a person estimating the mass of an object by handling it almost invariably uses a property of the object we do not study in this course. He "hefts" it by moving his hand up and down, to sense its inertia.)

45 **Two astronauts on the moon use an equal-arm balance to find the mass of a specimen of moon rock; the specimen has a mass of 35.83 grams. When the astronauts return to earth and mass the specimen once again, will they find that the mass of the rock on earth is more than, equal to, or less than the 35.83 grams they measured on the moon?**

They will find that the mass of the rock is still 35.83 grams on earth. The mass of the rock is the measure of the amount of matter in it and is independent of where it is located. [Some students may be confused by hearing that objects *weigh* less on the moon; this is true, but relates to a different kind of a measurement than that made with an equal-arm balance, one that we have not used in this course.]

Characteristic Properties

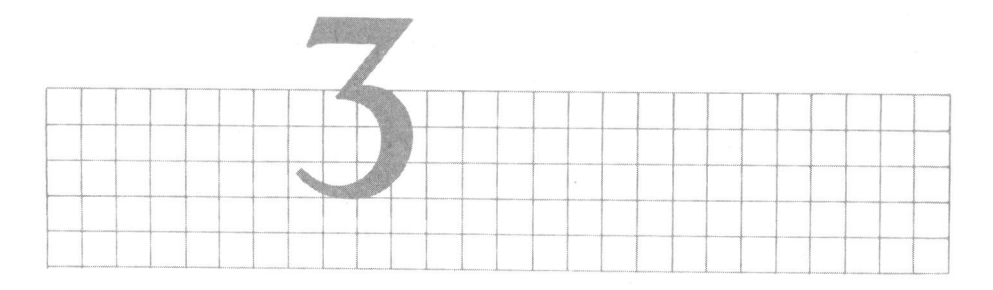

3

Overview of the Chapter

The selection of characteristic properties for discussion in this chapter and in Chap. 4 has been guided by one consideration: the usefulness of the characteristic property in separating mixtures. Hence we concentrate on freezing point, boiling point, and density in this chapter, and on solubility in Chap. 4.

Other characteristic properties that are relevant to the atomic model of matter, such as the compressibility of gases and thermal expansion, are introduced in Chap. 9.

The suggested schedule for the chapter is:

Sections 1-4 (three expts., probs. 1-2; 25; 33)	6 periods
Sections 5-6 (no expt., probs. 3-6; 26-30)	1 period
Sections 7-9 (three expts., probs. 7-17; 31; 34-35)	6 periods
Sections 10-11 (no expt., probs. 18-24; 32)	1 period
Total	14 periods

PROPERTIES OF SUBSTANCES AND PROPERTIES OF OBJECTS 3.1

"How do we know when two substances are different?" Because the emphasis is on the substance and not on a given sample, we have to look for properties that are independent of the size or the shape of the sample. These are characteristic properties of the substance.

EXPERIMENT: FREEZING AND MELTING 3.2

The freezing point of a substance is identified as the temperature at which a plateau occurs on the cooling curve when a substance changes from a liquid to a solid. To show that the freezing point does not depend on the amount of substance present and is, therefore, a characteristic property, have each group determine the freezing point of a different amount of material.

44 Characteristic Properties

The usefulness of the freezing point as an identifying characteristic property is illustrated by giving the class two different substances and allowing them to discover that the substances are different because they have different freezing points.

Typical graphs are shown in Figs. I and II. Both graphs show a slight dip just before the substance starts freezing. Many substances exhibit this effect. The liquid cools off below its freezing point; then, as soon as crystals begin to form, the temperature rises a little and remains constant until all the liquid has solidified. The dip in the graph is not always observed in doing this experiment and, therefore, may not appear on the graphs your students plot. Do not make a great point of it. It is not important in this experiment.

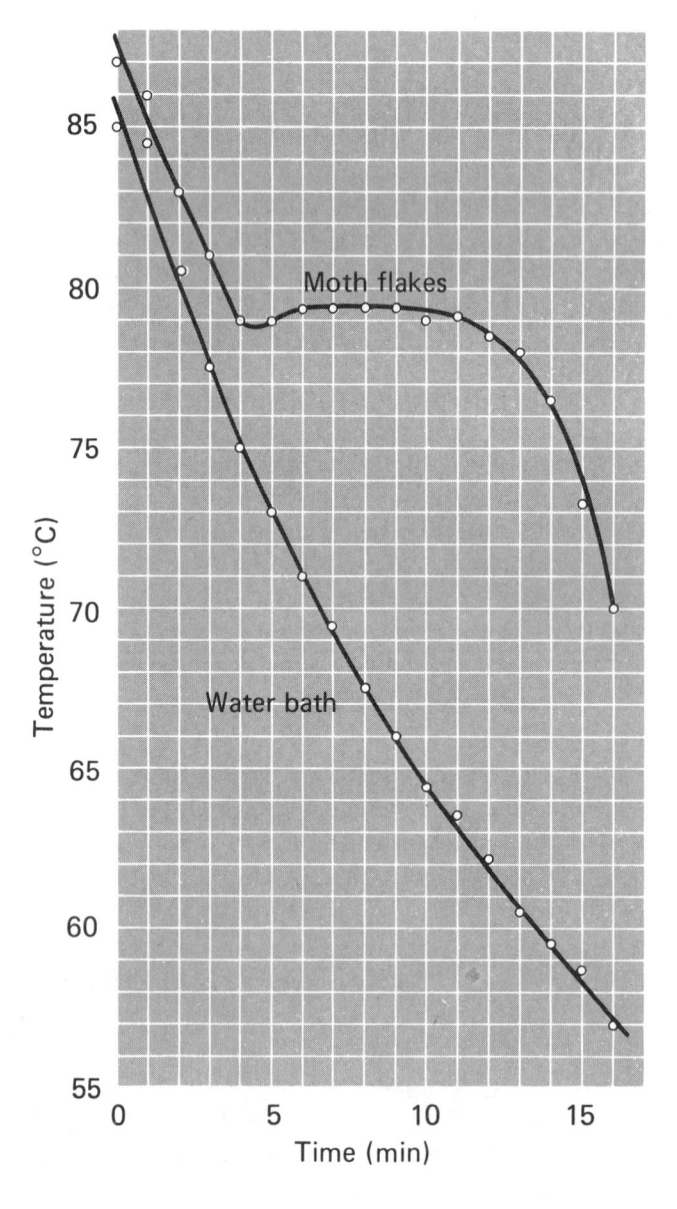

Figure I

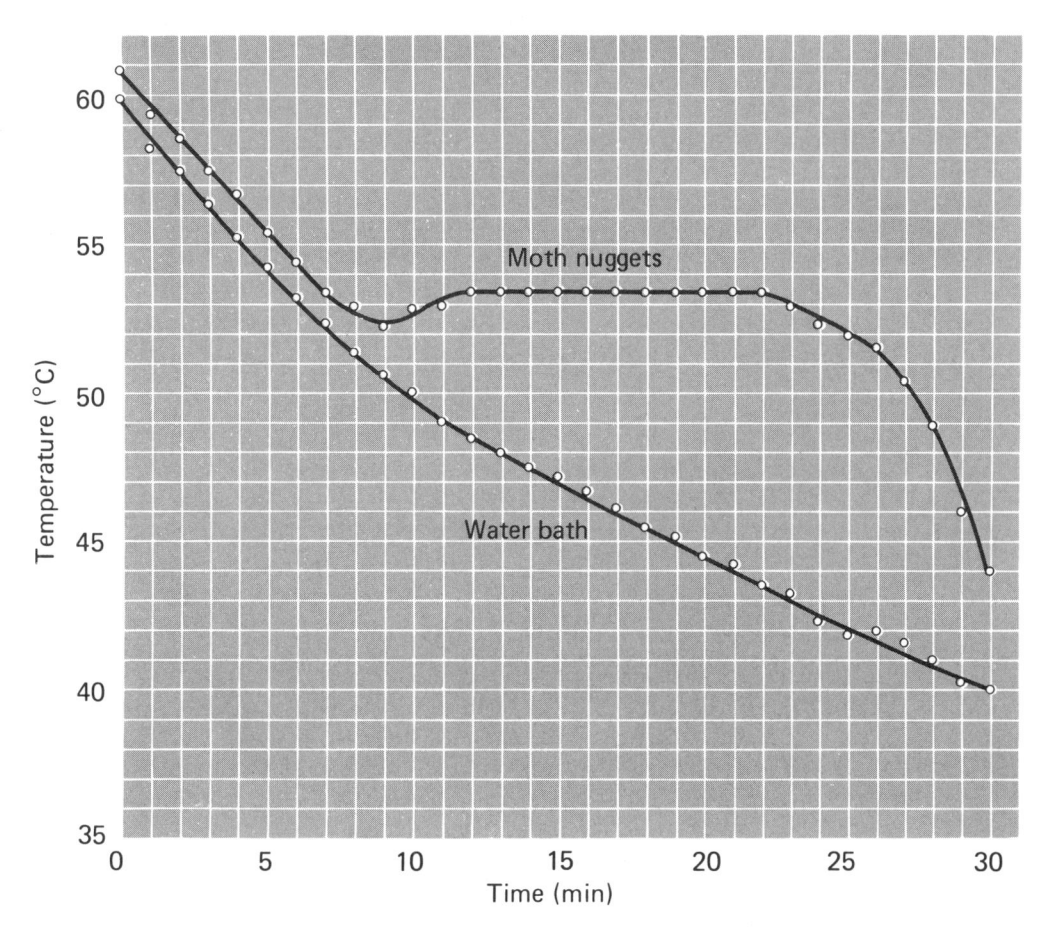

Figure II

We are not concerned here with the question of why the temperature remains constant while the substance freezes. It is sufficient that students realize that the plateau does give the freezing (or melting) point, and that the substance does solidify during the period when the temperature remains at a constant level.

Do not bother about explaining why the two curves have the shapes they do. This experiment is concerned only with the identification of freezing point as a characteristic property, not with the rate at which substances cool.

The Experiment

Slit cork stoppers supporting the thermometers make the whole temperature scale visible. They are made by boring a hole lengthwise through the stopper and then cutting out a wedge-shaped section. The thermometer should be fixed in the liquid so that the bulb is clear of the bottom and centered in the test tube.

To determine the extent of the variation in calibration of the thermometers, you can check them all together in a single large beaker of boiling water. It is good practice to label each thermometer with a number on a piece of tape or a tag.

46 Characteristic Properties

By plotting the cooling curve of both the water bath and the substance under investigation, students will see that the plateau is characteristic of the freezing substance and not of the water in the water bath as it cools to room temperature.

Give half the class varying amounts of moth flakes (naphthalene) and the other half moth nuggets (paradichlorobenzene) without telling them that they are different substances. At this time point out to your students that they do not all have the same amount of material.

In order to get the experiment completed in a single class period, you may have to have the apparatus set up and hot water in the water bath before the class period begins. Use two burners to speed up the melting of the material. A tube containing more than 10 g of paradichlorobenzene or a beaker larger than 250 cm^3 for the water bath will extend the experiment beyond a class period. To speed up the cooling, see that the water in the beaker comes just above the level of the material in the test tube, and remind your students to stir the water during the cooling.

Be sure that all the material is molten when readings are started. Your students should take readings until the temperature has again started to fall after remaining constant during the freezing.

CAUTION: When the experiment is finished, be sure that students melt the material in the test tube before extracting the thermometer, so that they will not break it. This can be done directly over the burner flame, but care must be taken not to heat the material beyond the range of the thermometer. After the thermometer has been withdrawn, the material in the test tube can be allowed to freeze, and the test tube of solid can be stored for use another time. The material should not be disposed of in the sink. The thermometers and test tubes can be cleaned with alcohol, which dissolves both moth flakes and moth nuggets.

The student should use the same thermometer for Expts. 3.2 and 3.3 to avoid apparent differences in melting point resulting from small differences in thermometer calibration.

Drawing Graphs

When your students draw their first graphs there should be a class discussion of the following.

Before drawing the two axes of a graph and marking off the scale divisions on them, you must give careful thought to the choice of scales. The two axes should both be as long as they can be and still allow convenient scales to be chosen along them. Students just learning to draw graphs often proceed without careful thought and compress one or the other of the scales to the point where the whole graph occupies only a narrow strip of the graph paper. Each axis should be labeled with the name and units of the quantity being plotted.

The scales, as a general rule, should start from zero. Sometimes, however, depending on what is being plotted, the origin of one or both scales may be other than zero (for example, see Fig. 3.2). One learns what to do only from experience. The scales should read with increasing values toward the right and toward the top.

A convenient scale is one in which (1) the highest value to be graphed fits on the axis and (2) decimal multiples (such as 100, 10, 1, or

0.1) of the unit being graphed can be easily located along the axis. A scale of time in minutes, for example, can be chosen with 10 spaces for 1 min, or five spaces, or two spaces, or one space; however, such numbers as 13, 11, 9, and 7 spaces are useless in plotting decimal numbers. A scale of four spaces is of doubtful use, and three is a poor choice.

The axes should be marked off with *equally* spaced numbers that relate in a simple way to the decimal system, such as 1,2,3,4 ... or 2,4,6,8, ... or 5,10,15,20, ... , for example. If students make scales with successive numbers such as 13,26,39,52, ... you can ask them how they would like to use a centimeter scale marked off in this manner! It is wise to have a number on the scale or a heavy line at least every five spaces. Too many spaces between numbers make it difficult to locate a particular space. However, if the numbers are unnecessarily crowded, close together, they will be hard to read and it will be difficult to distinguish the value corresponding to each point when the graph is read.

Points can be small crosses or circles made with a sharp pencil. (Squares or other symbols can be used when two or more curves are plotted on the same set of axes.) Points should be large enough not to be obscured by the curve that goes through them. Insist from the beginning that your students estimate to $\frac{1}{10}$ of the smallest division on their graph paper both in locating points and in reading graphs. In general, curves should be smooth and not broken lines. (Graph lines describing functions are usually called "curves" whether they are curved or straight.) The "best" curve is drawn by trial and error to "split the middle" of the pattern of data points if these are somewhat scattered because of experimental errors. A good graph, like good writing, should be reasonably neat and include everything needed for clearness, but should not be cluttered with unnecessary information that makes it difficult to read.

The preceding remarks on plotting graphs can serve as a general guide. There are exceptions to nearly every suggestion, but these can be dealt with when they arise. If students draw graphs resembling those in the text they will be off to a good start.

———   ———   ———

To summarize the results effectively, list on the chalkboard the mass of material used and the plateau temperatures found by your students. Make a histogram of the plateau temperatures.

Answers to Questions

All the graphs have flat sections.

All the flat sections of the graphs occur at two temperatures. These do not depend on the amount used.

The fact that all the graphs show a temperature close to either 80°C or 53°C for the flat section indicates that two different substances were investigated although other properties such as color and smell were the same. Since the readings of two thermometers may differ slightly from each other, there will be some difference in the recorded melting points for each substance. The histogram in Fig. III makes it clear that thermometer differences are small compared with the differences in the melting points of the two substances.

48   Characteristic Properties

Moth nuggets   Moth flakes

[Histogram with two distributions: one centered near 50–55°C (Moth nuggets) and one centered near 75–80°C (Moth flakes), x-axis labeled from 40 to 85]

Melting point (°C)

Figure III

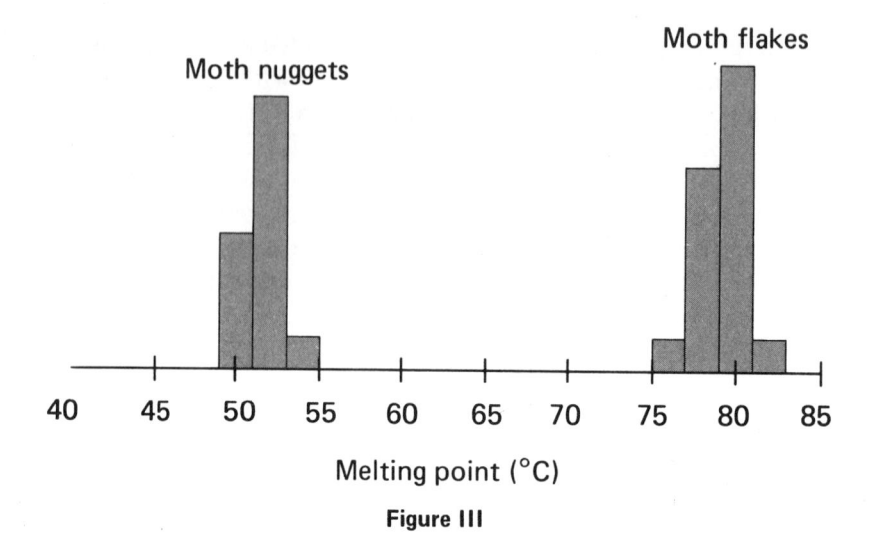

Apparatus and Materials

Pegboard
2 Burners
Burner stand
Test tube (20 × 150 mm)
2 Clamps
2 Slit cork stoppers
Beaker (250 cm^3)
2 Thermometers

5 to 15 g moth flakes (naphthalene) *or*
5 to 15 g moth nuggets (paradichlorobenzene) small enough to fit in test tube
Matches
Water
Paper towel
Stirring rod
Safety glasses

3.3 EXPERIMENT: MICRO-MELTING POINT

In the previous experiment your students conclude that the freezing point is independent of the amount of substance over a mass range of about 3 to 1 (5 to 15 g samples). In addition they are told that a plateau occurs at the same temperature, whether it is determined by freezing or melting a substance. In this experiment the melting point of about 10^{-4} of the mass of the same substances utilized in Expt. 3.2 is determined.

The Experiment

The glass threads remaining on the two pieces of capillary after they have been separated should be broken off close to the hollow part of the tube before sealing. (Some time will be saved if melting-point tubes—capillary tubes closed at one end—are purchased commercially.)

It is best to have only one or two tiny crystals in the tube. Then one can easily see when they melt. Give your students the same substance they used in Expt. 3.2 so they will know at about what temperature they should expect it to melt. As the melting point is approached in the water bath, the water should be heated very slowly (about 2°C/min) and the tiny crystals watched continuously. The thermometer must be read at the instant the crystals melt.

After the technique has been mastered, a second attempt, using the other piece of capillary, will probably give more accurate results. If there is time, let the student determine the melting point of an unknown to find out if it could be the same substance as used for the first trial. Acetamide (81°C) and a stearic acid (70°C) are other possible low-melting-point solids, but a solid that melts above the boiling point of water will serve just as well for the comparison, provided you have thermometers that read to sufficiently high temperatures and if cooking oil is used in place of water in the water bath.

The following sources of errors should be eliminated: (1) different thermometers used in the two experiments; (2) too fast heating of the water, so that its temperature and that of the inside of the capillary are not the same; and (3) failure to read the thermometer soon enough. The last two errors are the most common, and either will produce a melting point a degree or two higher than the freezing point of Expt. 3.2. Because of this, class results may be consistently higher than the results of Expt. 3.2. Nevertheless, results will indicate that the freezing and melting points are very close, and an improvement in technique will show that they are the same.

Answers to Questions

The observed melting point is within a degree or two of the freezing point observed in Expt. 3.2.

The crystals are about the size of cubes a tenth of a centimeter on a side. Their volume is therefore about 0.1 cm $\times$ 0.1 cm $\times$ 0.1 cm = 10^{-3} cm^3. The volume of material melted in the test tube in Expt. 3.2 was about 10 cm^3. Thus the volume, and hence the mass of material, in Expt. 3.2 was about 10 cm^3/10^{-3} cm^3 = 10^4 greater.

(You may find that students rebel at such rough calculations. You can point out that rough estimates to order of magnitude, the nearest power of ten, are often very useful, particularly in this case where an order-of-magnitude calculation is all that is needed to make the point—that melting point is a characteristic property independent of the mass over a wide range.)

Apparatus and Materials

Pegboard
Clamp
Thermometer
Slit cork stopper
Beaker
Burner
Capillary tube, thin-walled, 1.5 to 2.0 mm diameter
2 Small rubber bands
Burner stand

Small piece of moth nuggets (paradichlorobenzene) *or* small piece of moth flakes (naphthalene)
Matches
Water
Paper towels
Stirring rod
Safety glasses

3.4 EXPERIMENT: BOILING POINT

In this experiment, the boiling points of two liquids are investigated. Choose two of the following: water, methanol, or burner fuel. Do not have the class use all three; you would have too little data on any one case to make firm conclusions.

If you dispense the two liquids from four identical bottles without labels, you can help the class from jumping to the premature conclusion that there are two and only two liquids being investigated. The students may notice the faint odor of methanol or the stronger odor of burner fuel and thus realize in advance that there are at least two different liquids. However, that there are only two liquids should not become clear until the post-laboratory discussion.

The Experiment

Divide your class into four groups and distribute the samples as follows:

Group 1: ¼ test tube of liquid 1
Group 2: ½ test tube of liquid 1
Group 3: ¼ test tube of liquid 2
Group 4: ½ test tube of liquid 2

This is the first time your students will be heating flammable liquids. *Be sure that strict safety procedures are followed.* Check all setups for leaks and blockages. There are good reasons for the equipment setup pictured in Fig. 3.4 in the text: The burner stand distributes the heat of the burner more evenly and prevents too rapid boiling or "bumping." The pegboard is placed vertically to allow sufficient height for the placement of clamps on the test tube and to reduce air currents. The cold water bath condenses the boiled liquid and prevents vapor from being released into the room.

Boiling chips must be used. The thermometer bulb should be completely immersed in the liquid so as to determine the boiling point of the liquid and not the temperature of the condensing vapor. The heat of the flame should be sufficient to maintain a steady flow of vapor into the condensing tube, but not so great that too vigorous boiling will force liquid into the delivery tube. If this seems likely to happen, the burner can be moved slightly to the side to reduce the amount of heat reaching the test tube. Boiling should be continued until roughly one-fifth of the liquid has been boiled away.

After the graphs have been plotted, a histogram can be constructed using the temperature scale along the horizontal axis of the histogram.

Typical experimental data have been plotted in Figs. IV and V.

Characteristic Properties 51

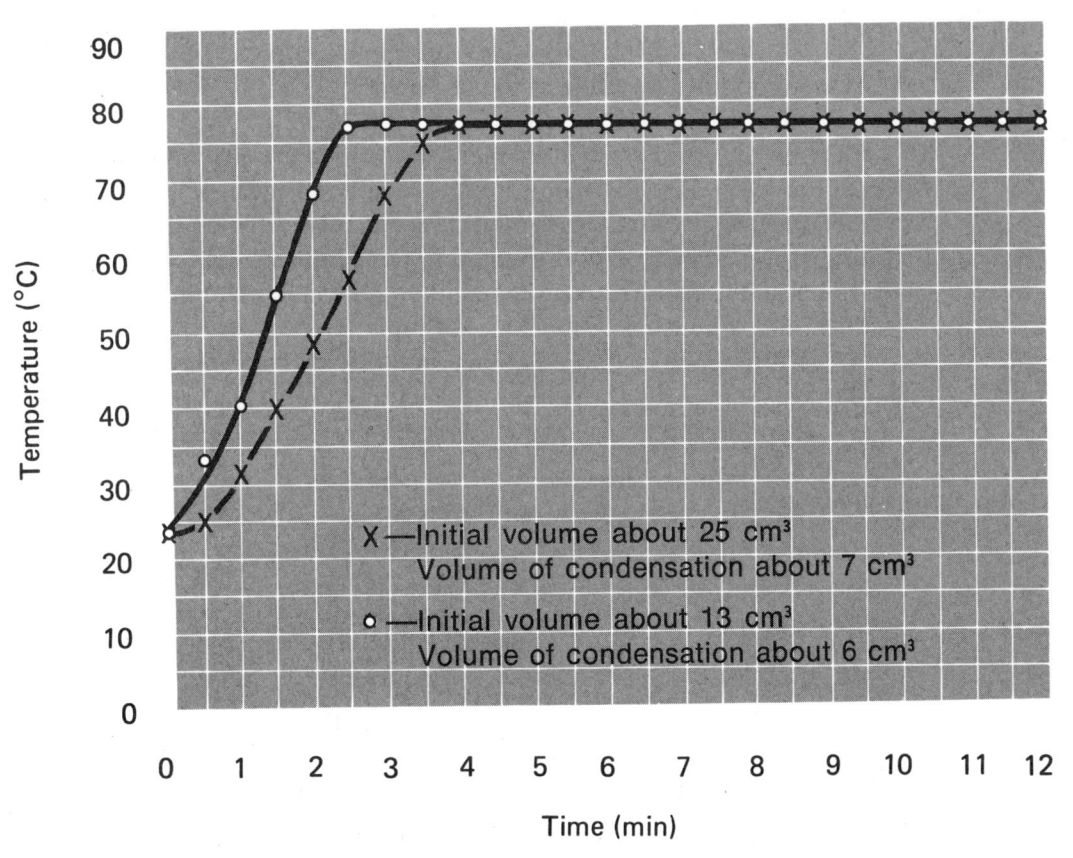

Figure IV Time-temperature graph for burner fuel.

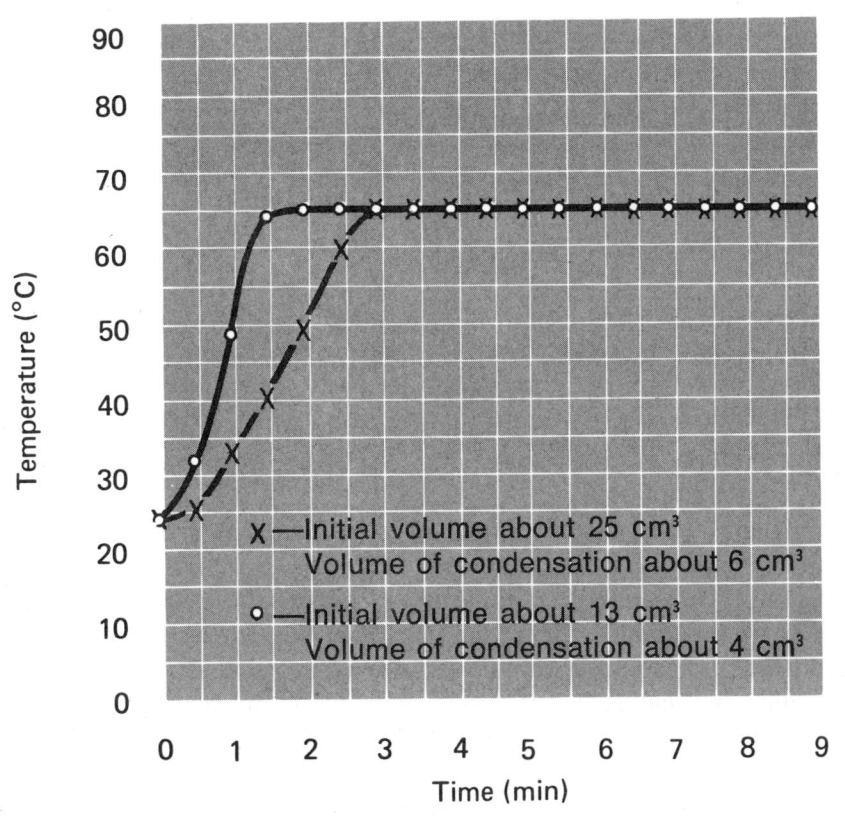

X—Initial volume about 25 cm³
Volume of condensation about 7 cm³

○—Initial volume about 13 cm³
Volume of condensation about 6 cm³

Figure V Time-temperature graph for methanol.

X—Initial volume about 25 cm³
Volume of condensation about 6 cm³

○—Initial volume about 13 cm³
Volume of condensation about 4 cm³

Characteristic Properties

Answers to Questions

The graphs do not look alike at the beginning.

All the graphs have a flat section.

Once the liquid started boiling, the temperatures in about half of the test tubes were close to one of the following temperatures: Water, 100°C; Burner fuel, 78°C; Methanol, 65°C; the temperatures in the rest of the test tubes were close to another of these temperatures.

The boiling points do not depend on how much liquid there is at the start or on how much is left.

The difference in boiling points reveals that there were two different liquids.

The boiling point is a characteristic property of the kind of liquid.

Apparatus and Materials

Pegboard	Paper towel
Burner	Large clamp, for large test tube
Burner stand	Small clamp, for small test tube
Large test tube	Short bend of glass tubing
Small test tube	Rubber or plastic tubing (25 cm)
Two-hole rubber stopper	Beaker (250 cm^3)
(No. 4 with one hole large enough to hold a thermometer)	Two of the following: water, burner fuel, methanol
Thermometer	Graph paper
Boiling chips	Glycerin
Matches	Safety glasses

One hole of a two-hole rubber stopper can be enlarged with a file or drill so that it will hold a thermometer.

3.5 DENSITY

Students will find the units of density more understandable if the units are referred to as mass of a unit volume (in this case 1 cm^3) instead of mass divided by volume. This usage stresses the physical significance of density and not just the mathematical "ratio." Density is not a ratio; ratios are obtained only when similar things are compared, such as two masses or two volumes.

Actually, your students are already familiar with a similar situation from daily experience. What you pay for several pounds of butter (the cost) is not a characteristic property of any particular brand, because it depends on the number of pounds purchased. For example, knowing that the butter costs $4.00 does not tell us what brand it is. But the cost of one pound is a characteristic property of the brand (excluding quantity discounts, of course). The cost of one pound, known as the price, has actually the complex units of cents/lb. If we know the price per pound of a given brand of butter and the amount of money we have, we can find the number of pounds we can purchase; or if we know the price and the number of pounds we wish to purchase, we can find the amount of money we need. The relation between amount of money, number of pounds, and price is quite analogous to that between mass, volume, and density. We can look upon a price as a kind of "density of cost."

To strengthen the idea that density is independent of mass and volume, you can break a piece of chalk in half and point out that each smaller piece has half the mass and half the volume of the original piece, and so the mass divided by the volume is the same for the smaller pieces as for the original larger piece.

DIVIDING AND MULTIPLYING MEASURED NUMBERS 3.6

Calculating quotients and products to too many insignificant digits is widespread. The pocket calculator aggravates the situation by taking the tedium out of the calculation.

The purpose of this section is twofold: to teach students who calculate on paper where to stop, and to teach those who use a calculator what to discard. Read it in class and see to it that your students apply what they learn in all future calculations.

Students without a hand calculator or a slide rule will save themselves much time and useless labor by dropping off digits that are not significant after *each* successive multiplication or division. This will be immediately apparent in the next section, Expt. 3.7: The Density of Solids, where the product of three lengths is calculated.

Considering the age level of the students and their probable level of mathematical competance, the rule of thumb given in the text is the simplest procedure to follow in deciding on the number of digits to keep. There are more sophisticated ways of handling the results of computation, but they are too long and complex for application in this course.

The subject of significant digits is probably new to your students. In the past it is likely that they have been required to retain every digit, or to round off in terms of decimal places. In fact, the number of decimal places is an accident of the units selected to express a number and has nothing to do with accuracy or precision. For example, the volume of a small object can be expressed equally well as 1.95 cm^3 or as 0.00000195 m^3. Both numbers, having the same number of significant digits, express the volume to the same accuracy.

EXPERIMENT: THE DENSITY OF SOLIDS 3.7

This experiment is far longer and more difficult for students than might appear at first sight. Because this experiment involves more measurements and calculations than do those in Chap. 2, you will need to pay close attention to the organization of the students' notebooks.

Allow enough time for student preparation, for performing the experiment, and for post-laboratory group analysis; it will pay dividends not only for this particular experiment but also for many of those to come.

The first part of the experiment is designed to sharpen the students' understanding of the difference between mass and volume; the two cubes of equal volume obviously have different masses.

Guesses as to the mass and density of a third object of different shape and volume prove futile, and without careful measurement the substance of which it is made cannot be identified as that of either of the small cubes. Careful measurements are needed; in making them, the students reinforce their understanding of volume and density measurement for a simple rectangular solid.

In the last part of the experiment your students make use of what they learned in Expt. 2.3, Measuring Volume by Displacement of Water, to determine the density of an irregularly shaped stone.

The Experiment

Care must be taken in measuring the dimensions of the small cubes. It cannot be assumed that they are perfect cubes, and all three dimensions must be measured. In making the measurements, the cubes must be turned in such a way that a new dimension is measured each time, rather than the same one on a different face. One way is to turn the cube left or right for the second measurement, and then up or down, followed by a second left or right turn for the third measurement; another is to start at a corner and go three different ways.

The dimensions of the objects are measured with a centimeter rule. Insist that your students estimate the dimensions to a fraction of a millimeter, as discussed in Sec. 2.2.

In calculating the densities of the cubes and the slab, your students will probably have to do a good deal of careful checking to ensure error-free computations. Be consistent in requesting that they follow the procedures specified in Sec. 3.6 in expressing the results of computations.

The histograms for the densities of the two cubes and the slab reproduced in Fig. VI from a pilot class clearly differentiate two substances, and you can now show your students that the most probable densities are 2.8 g/cm^3 and 7.8 g/cm^3. Two digits are all that are necessary here. Note that by including 0 g/cm^3 on the histogram, the relative spread of the data is clearly shown.

Sample Data

The dimensions of the objects (from one manufacturer):

Object	Length (cm)	Width (cm)	Height (cm)
Cube No. 1	1.25	1.25	1.25
Cube No. 2	1.25	1.25	1.25
Third object	7.60	2.50	0.60

The volume, mass, and density of the objects:

	Volume (cm³)	Mass (g)	Density (g/cm³)
Cube No. 1	1.95	15.14	7.76
Cube No. 2	1.95	5.26	2.70
Third object	11.4	30.85	2.71

The density of a stone:

Mass of stone	5.19 g
Initial graduated-cylinder reading	27.2 cm³
Final graduated-cylinder reading	29.0 cm³
Density = 5.19 g/1.8 cm³ =	2.9 g/cm³

Figure VI

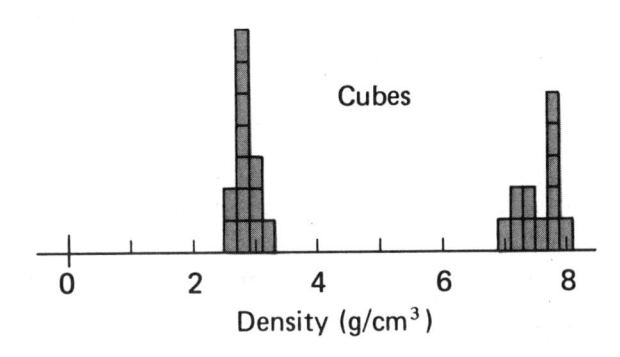

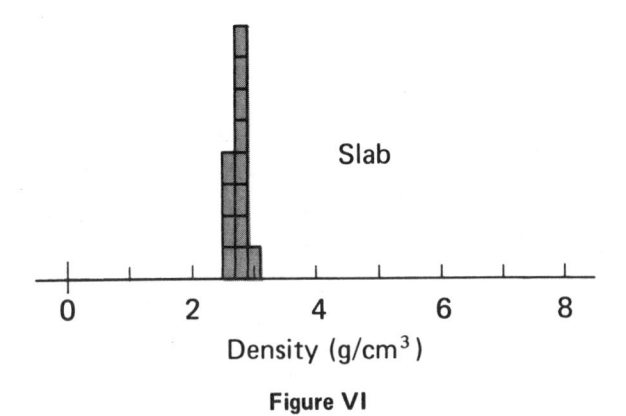

Answers to Questions

The two cubes unquestionably have different masses.

The masses of the cubes are 15.14 g for one and 5.26 g for the other. The cube with the greater mass has the greater density, since the volumes are nearly the same.

It is almost impossible to compare the mass of the third object with the mass of each of the cubes just by handling them. The third object seems to be intermediate in mass between the two cubes.

The density of the third object is even more difficult to guess than its mass, and it is impossible to decide whether or not it is made of the same substance as one of the cubes.

56   Characteristic Properties

As seen from the data, the third object could be the same substance as cube No. 2, but one cannot be sure. (Two different substances can have the same density, as your students will soon learn. One must compare other characteristic properties in order to be sure.)

Samples from the same rock may have different densities if they differ in porosity or in the proportion of different minerals making up the rock, but such differences are not likely to be greater than errors due to the limitations of measuring mass and volume.

Apparatus and Materials

Set of 3 metal blocks
Balance
Centimeter rule
Graduated cylinder (50 cm^3)

Stone small enough to fit into graduated cylinder
Water
Paper towel

3.8   EXPERIMENT: THE DENSITY OF LIQUIDS

The two samples of liquid investigated are different concentrations of Epsom salt and water. Both solutions are transparent and odorless, and cannot be distinguished by inspection. Their densities, however, differ sufficiently so that they can be distinguished by simple density measurements. The actual densities are not important. It should be noted that the density of a solution will change with the concentration of the solute, but there is no need to mention this to your students.

The Experiment

The use of a histogram in discussing the class results in this experiment is vitally important. Small errors in measurement or technique by any single group may cause the results to be too close together to distinguish the two liquids. When a histogram of all the results is drawn (Fig. VII,) two distinct peaks become evident, even though the distribution may overlap.

It is more accurate to mass the liquid in a container before pouring it into the graduated cylinder. The mass of the liquid whose volume is measured is the mass of container and liquid minus the mass of container after the liquid is poured into the graduated cylinder. If the volume is determined first and the liquid poured into a previously massed container, the mass of the liquid that sticks to the sides of the graduated cylinder is not measured; i.e., the sticking liquid is included in the volume measurement but not in the mass measurement. (This question is raised to bring students' attention to the need for careful thought about how a measurement should be made.) Often, as is the case in this experiment, the order in which a procedure is carried out may be important. Your students may suggest determining the density by measuring the mass and volume in the graduated cylinder. A 10-cm^3 graduated cylinder is convenient for this and can be held on the balance with a rubber band and paper clip, like the test tube in Fig. 3.5. However, it is still worthwhile to discuss, in the pre-lab, the logic of the problem in the text.

Sample Data

	Solution I	Solution II
Mass of container and liquid	20.54 g	17.59 g
Mass of container	9.12 g	7.22 g
Mass of liquid	11.42 g	10.37 g
Volume of liquid	9.7 cm^3	9.7 cm^3
Density	1.18 g/cm^3	1.07 g/cm^3

Figure VII

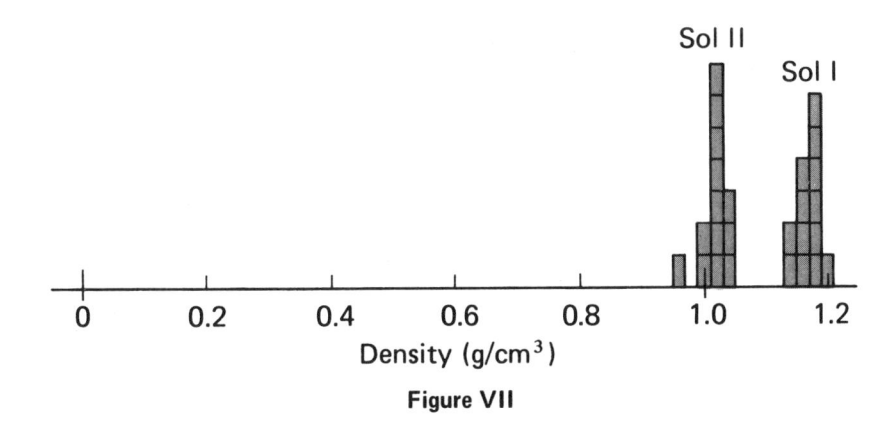

Answers to Questions

The two liquids cannot be told apart just by inspection and smelling. (Students are told not to taste them; this is a general rule. You may well emphasize in class that *no substance* in the laboratory is ever to be tasted and that unknown substances are to be smelled cautiously.)

The liquids are different, since the class histogram shows that their densities differ by more than the experimental error.

Apparatus and Materials

Small container (about 15 cm^3)
Graduated cylinder (10 cm^3)
Balance
Solution I (10 cm^3)
Solution II (10 cm^3)
Paper towel

Solution I is prepared by dissolving 250 g of Epsom salt (MgSO$_4 \cdot$ 7H$_2$O) in 500 cm^3 of water. Solution II is prepared by dissolving 50 g of Epsom salt in 500 cm^3 of water.

EXPERIMENT: THE DENSITY OF A GAS 3.9

One of the purposes of this experiment is to see that the density of a gas is about 10^{-3} times that of liquids and solids. This will become important in Chap. 9. In Expt. 2.13 students found that a gas has mass. In this experiment they will use the same method as in Expt. 2.13 to produce a sample of gas and measure its mass. However, they will now collect the gas over water, measure its volume, and then calculate its density.

There are many steps in the procedure that must be done carefully and quickly. Careful pre-lab instruction is essential, including a demonstration of the apparatus and a sample run by your students without taking data. Following are some suggestions for pre-lab discussion:

The method of collecting the gas and the determination of the volume of gas. The easiest way to determine the volume is to measure the volume of water required to refill the bottle.

The justification for stopping before all the gas is evolved. Gas is evolved for about an hour from one tablet. However, at least 95 percent of the gas is evolved within 5 minutes. It is suggested, therefore, that the experiment be run for 10 minutes and that the reason for this be discussed in the pre-lab.

All massings must be performed as accurately as possible, because the change in mass of the test tube and contents is small. There should be no large air bubbles in the collecting bottle when the students insert the rubber tube to collect the gas. The rubber tube should extend to the top of the collecting bottle to reduce the amount of gas that dissolves in the water. (The gas is carbon dioxide, but this is irrelevant at this time.) If the tube goes only into the bottom of the bottle, a considerable quantity of gas will dissolve from the bubbles as they rise to the top. Thus the apparent volume of gas will be too small, and the calculated density will then be too high. (See probl. 17.)

CAUTION: The glass tubing should not extend much beyond the stopper into the generating tube. The delivery tube must be withdrawn from the collecting bottle *before* the test tube is unstoppered. This prevents air from running through the delivery tube into the collecting bottle if the water level in the bottle is above the water level in the pan, or vice versa.

The task of inverting the collecting bottle full of water can be simplified if a piece of wet paper towel is placed over the mouth of the bottle. If the piece is about twice as large as the opening and there are no air spaces, the bottle can be inverted and placed in the bucket without spillage.

There may be some variation in Alka-Seltzer tablets; however, a 500-ml collecting bottle should be more than adequate.

This is not an easy experiment for students to do accurately. After class results are tabulated on the chalkboard and sources of error are discussed, it may be worthwhile to have the whole class repeat the experiment the next day. The accepted value of the density should not be given to the students, but even without knowing this figure they are very likely to get better results the second time.

Sample Data

Initial mass of test tube + water + tablet	35.00 g
Final mass of above	34.22 g
Mass of gas produced 35.00 g − 34.22 g =	0.78 g

Here we are forced to do something that experimenters try to avoid: to use a small number that is the difference of two large ones. Even a small fractional error in either of the large numbers will cause a large fractional error in the small one. Here, for example, an error of 2 percent in 35.00 g would yield an answer of 0.78 ± 0.7 g for the mass of the gas, which renders the experiment almost useless.

Volume of gas   411 cm^3
Density of gas = 0.78 g/411 cm^3 = 1.9 × 10^{-3} g/cm^3

The histogram in Fig. VIII shows the results of 16 trials of the experiment. From the variation in the values for density, the best value and also the range of experimental error can easily be determined.

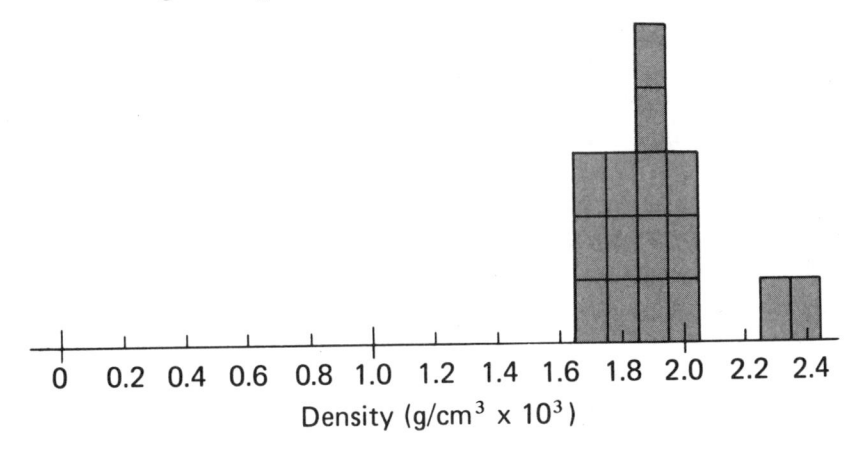

Figure VIII

Density (g/cm^3 × 10^3)

The two trials that gave values of 2.3 × 10^{-3} and 2.4 × 10^{-3} g/cm^3 were done with the gas delivery near the bottom of the collecting bottle instead of near the top, where it should have been. As we can expect, these values are too high as a result of gas dissolving in the water in the collecting bottle. The other results indicate that the most probable value that can be obtained with this apparatus is close to 1.9 × 10^{-3} g/cm^3.

The spread of results between 1.7 × 10^{-3} g/cm^3 and 2.0 × 10^{-3} g/cm^3 shows that in using this method you can expect experimental errors that may lead to an error in the results of about ±0.2 × 10^{-3} g/cm^3 or 10 percent. It is likely that the spread of values obtained by your class will be larger than that shown in the graph. Nevertheless, the density of a gas can be measured, and it can be seen that the density of this gas is about one thousandth of the density of a typical liquid.

Answers to Questions

When the bottle of gas is removed from the pan of water, no water should be allowed to run out of the bottle (and be replaced by air) if an accurate determination is to be made of the volume of gas collected.

The volume of water displaced by the gas equals the volume of the gas produced and collected.

The mass of the test tube, with its contents, stopper, etc., is less than before because the gas that was produced escaped from the test tube. (Some students may wish to try to identify the gas by using Table 3.1. This is fine, but do not tell them that the gas is carbon dioxide until you are completely finished with the experiment.)

The conservation of mass has been assumed in this experiment.

60 Characteristic Properties

Apparatus and Materials

Balance	Bottle (at least 500 cm^3)
Clamp	18-in. rubber or plastic tubing
Test tube	Dish pan or bucket
Right-angle glass bend	Water
No. 2 one-hole rubber stopper	Graduated cylinder
Alka-Seltzer tablet	Safety glasses

3.10 THE RANGE OF DENSITY

Table 3.1 is worth discussing in class. Specifically point out the range of densities of liquids and solids, and compare this in magnitude with the range of density of gases.

Also mention qualitatively the fact that an object will rise through a liquid if the density of the object is less than the density of the liquid. A less dense gas will rise in a denser gas; hydrogen rises in air. Do not get involved with the quantitative aspects of buoyancy.

3.11 IDENTIFYING SUBSTANCES

For good readers you may call their attention to the Table 3.2 and assign the section for reading at home, and then discuss in class. For poorer readers it may be better to discuss Table 3.2 in class, and use the text for review.

CHAPTER 3—ANSWERS TO PROBLEMS

Sec.	Easy	Medium	Hard	Class Discussion	Home or Lab
1	1				
2-3	2†				
4	25	33		33	
5	3	4			
6	5, 6				
7	7†, 9†, 11†, 26	8, 10, 28, 29	27	8	
8		12, 30		12	31
9	13†	14, 15, 17, 34, 35	16		
10	18, 21†, 22†	19, 20, 32	23	23	
11		24			

1 In the following descriptions state which are properties of the substance and which are properties of the object.
   a) A sharp, heavy, shiny, stainless-steel knife
   b) A small chunk of black tar
   c) A beautifully carved wooden chair

Properties of the substance	Properties of the object
a) stainless-steel	sharp, shiny, heavy
b) black, tar	small
c) wooden	beautifully carved

2† The graph in Fig. A represents data from an experiment on the cooling of paradichlorobenzene. During which time intervals is there (a) only liquid, (b) only solid, and (c) both liquid and solid?

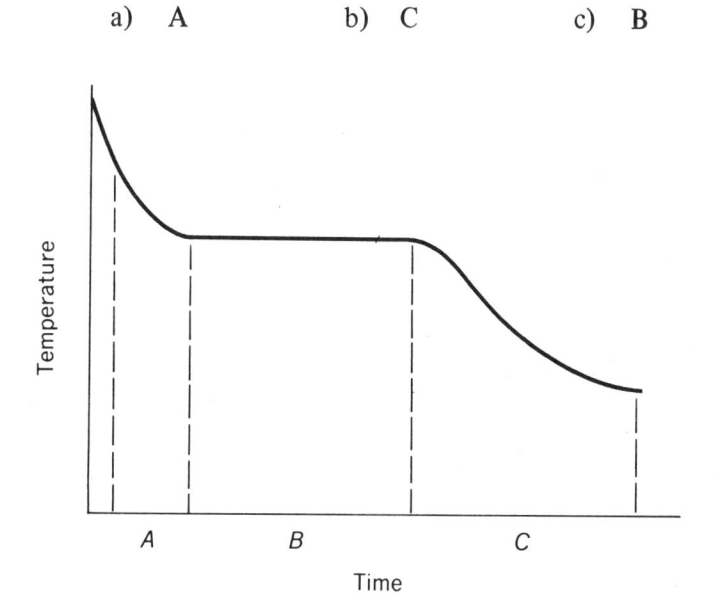

   a) A   b) C   c) B

Figure A

3 A parking lot is filled with automobiles.
a) Does the number of wheels in the lot depend upon the number of automobiles?
b) Does the number of wheels per automobile depend upon the number of automobiles?
c) Is the number of wheels per automobile a characteristic property of automobiles that distinguishes them from other vehicles?

The purpose of this question is to introduce the idea of a characteristic property that is independent of amount or number. Questions (a) and (b) are two of the very few questions in this course that can be answered by a simple "yes" or "no."
a) Yes.
b) No.
c) The number of wheels per automobile distinguishes automobiles from many other vehicles such as bicycles, motorcycles, etc., but not from all other vehicles; tractors, trucks, and wagons may also have four wheels.

4 a) Draw a graph of the number of wheels in a parking lot as a function of the number of cars.
b) Draw a graph of the number of wheels per car as a function of the number of cars in the lot.

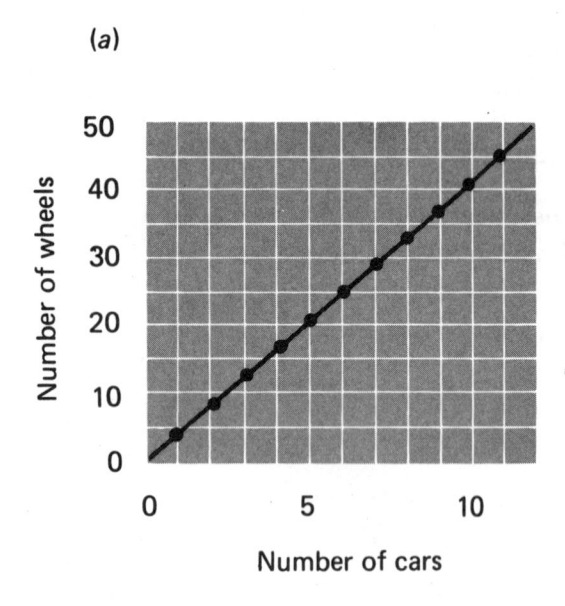

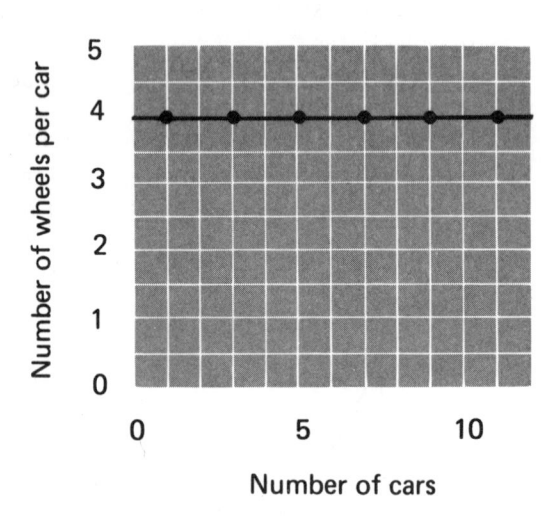

Figure IX

5 Calculate the following quotients to the proper number of digits.
a) $\dfrac{125}{23.7}$ c) $\dfrac{0.065}{32.5}$
b) $\dfrac{20.5}{51.0}$ d) $\dfrac{1.23}{0.72}$

a) 5.27   b) 0.402   c) 0.0020   d) 1.7

6 Calculate the following products to the proper number of digits.
   a) 4.72 × 0.52
   b) 6.3 × 10.08
   c) 1.55 × 2.61 × 5.3
   d) 3.01 × 5.00 × 25.62

   a) 2.5   b) 64   c) 21   d) 386

7† What measurements and what calculations would you make to find the density of the wood in a rectangular block?

Measure length, width, thickness, and mass of the rectangular blocks. Multiply length by width by thickness to get volume. Then divide the mass by the volume.

8 A student announced that she had made a sample of a new material that had a density of 0.85 g/cm^3. How large a sample had she made?

The size of the sample, whether measured by mass or by volume, cannot be determined from the density of the sample alone because the density of a substance is independent of the size of the sample. This is why the density is a characteristic property. The mass could be found only if the volume were also known, and vice versa.

9† A block of magnesium whose volume is 10.0 cm^3 has a mass of 17.0 g. What is the density of magnesium?

1.7 g/cm^3

10 Two cubes of the same size are made of iron and aluminum. How many times as heavy is the iron cube than the aluminum cube? (See Table 3.1, page 43.)

$$\frac{\text{Density of iron}}{\text{Density of aluminum}} = \frac{7.8 \text{ g/cm}^3}{2.7 \text{ g/cm}^3} = 2.9 \text{ times heavier}$$

11† a) A 10.0-cm^3 block of silver has a mass of 105 g. What is the density of silver?
    b) A 5.0-cm^3 block of rock salt has a mass of 10.7 g. What is the density of rock salt?
    c) A sample of alcohol amounting to 0.50 cm^3 has a mass of 0.41 g. What is its density?

    a) 10.5 g/cm^3
    b) 2.1 g/cm^3
    c) 0.82 g/cm^3

64   Characteristic Properties

12   You are given two clear, colorless liquids. You measure the densities of these liquids to see whether they are the same substance or different ones.
a)   What would you conclude if you found the densities to be 0.93 g/cm^3 and 0.79 g/cm^3?
b)   What would you conclude if you found the density of each liquid to be 0.81 g/cm^3?

a)   On the basis of the information given, the substances cannot be the same. However, the problem can lead to a discussion of the significance of the data given. Are the densities given really different, or are they the same within experimental error? If the experimental error were as much as ±0.07 g/cm^3 (about 8 percent), the true values could be 0.93 − 0.07 = 0.86 g/cm^3 and 0.79 + 0.07 = 0.86 g/cm^3, and both samples could be of the same substance. But if the experimental error were really so large, the densities should have been given as 0.9 and 0.8 g/cm^3 to indicate that the next digit is not known.
   If the experimental error were as low as 0.01 g/cm^3 (about 1 percent), then the true values could not be closer than 0.92 g/cm^3 and 0.80 g/cm^3, and we should be justified in deciding that the substances are different.
b)   In this case the substances may be the same. One cannot be sure, however, without checking other characteristic properties.

13†   A mixture of two white solids is placed in a test tube, and the mass of the tube and its contents is found to be 33.66 g. The tube is stoppered, and arrangements are made to collect any gas produced. When the tube is gently heated, a gas is given off, and its volume is found to be 470 cm^3. After the reaction, the mass of the test tube and its contents is found to be 33.16 g.
a)   What is the mass of the gas collected?
b)   What is the density of the gas collected?

a)   <u>0.50 g.</u>
b)   <u>1.1 × 10^{-3} g/cm^3.</u>

14   Experiment 3.9 is repeated with a sample of a different solid. Here are the data obtained:
Mass of solid, test tube, and water before action . . . . . . 35.40 g
Mass of test tube and contents after action . . . . 34.87 g
Volume of gas collected . . . . 480 cm^3
Could this gas be the same as that produced in Expt. 3.9?

No, the mass of the gas is 35.40 − 34.87 = 0.53 g; the density of the gas is $\frac{0.53}{480}$ = 1.1 × 10^{-3} g/cm^3. The density of the gas produced in Expt. 3.9 was about 1.9 × 10^{-3} g/cm^3. The densities of the two gases are too far apart to be the same gas.

Characteristic Properties 65

15 The volume of gas generated by treating 1.0 g of magnesium carbonate with 8.8 g of sulfuric acid is 200 cm³. The remaining acid and solid have a mass of 9.4 g. What is the density of the gas evolved?

Initial mass of materials is 8.8 g + 1.0 g = 9.8 g; final mass is 9.4 g, and the loss of mass is 9.8 g − 9.4 g = 0.4 g. This is the mass of the gas. The density is 0.4 g/200 cm³ = $\underline{2 \times 10^{-3} \text{ g/cm}^3}$.

16 If the volume of gas in the preceding problem is compressed to 50 cm³, what will the density of the gas now be? To what volume must the gas be compressed before it will reach a density of 1.0 g/cm³, a typical density of a liquid?

The density of the gas is now 0.4 g/50 cm³ = $\underline{8 \times 10^{-3} \text{ g/cm}^3}$. Volume = mass/density. The mass of the gas is still 0.4 g. The volume would be 0.4 g/(1.0 g/cm³) = $\underline{0.4 \text{ cm}^3}$.

17 The gas whose density you measured in the experiment of Sec. 3.9 dissolves slightly in water.
   a) How does this affect the volume of the gas you collect?
   b) How does this affect your determination of the density of the gas?

a) Some of the gas will dissolve in the water of the collecting system, and so the measured volume will be somewhat less than it should be.
b) The mass of the gas was determined by subtracting, at the end, the total mass of the test tube and its contents from the initial mass of the test tube, tablets, and water so that the mass measurement of the gas includes that which dissolved in the water of the collecting system. Therefore, the density as found by the student will be too high.

The gas that dissolves in the water in the *test tube* does not affect the density determination, because neither its mass nor its volume contributes to the calculations for the collected gas.

18 Write the following numbers in powers-of-10 notation.
   a) 0.001   0.1   0.0000001
   b) 1/100   1/10,000

a) 10^{-3} 10^{-1} 10^{-7}
b) 10^{-2} 10^{-4}

66 Characteristic Properties

19 Write each of the following numbers as a number between 1 and 10 times the appropriate power of 10.
   a) 0.006   0.000032   0.00000104
   b) 6,000,000   63,700

   a) 6×10^{-3} 3.2×10^{-5} 1.04×10^{-6}
   b) 6×10^{6} 6.37×10^{4}

   Question (b) is an extension that the student can work out by combining the information in the footnotes in Secs. 2.14 and 3.10.

20 Change the following numbers from powers-of-10 notation to ordinary notation.
   a) 10^{-2} 10^{-5} 3.7×10^{-4}
   b) 1.05×10^{-5} 3.71×10^{3}

   a) 0.01, or 1/100   0.00001, or 1/100,000   0.00037
   b) 0.0000105   3710

21† A small beaker contains 50 cm³ of liquid.
   a) If the liquid were methyl alcohol, what would be its mass?
   b) If the liquid were water, what would be its mass?

   a) 40 g.
   b) 50 g.

22† The densities in grams per cubic centimeter of various substances are listed below. Indicate which of the substances might be gas, liquid, or solid. (Refer to Table 3.1)
   (a) 0.0015   (b) 10.00   (c) 0.7   (d) 1.1   (e) 10^{-4}

   a) Gas   b) Solid   c) Solid, or liquid   d) Solid, or liquid   e) Gas

23 Estimate the mass of air in an otherwise empty room that is the size of your classroom.

   Accept any reasonable estimate of the size of the room. A typical classroom might be 7 meters wide, 10 meters long, and 3 meters high. The students will probably require some assistance in the arithmetic of numbers in scientific notation. The volume of the room would be 7 m × 10 m × 3 m = 210 m³. There are 100 cm × 100 cm × 100 cm = 1,000,000 = 10^{6} cm³ in one cubic meter (m³); therefore, the volume of the air in cm³ is 210 × 10^{6} = 2.1×10^{8} cm³. Since the density of the air is 1.2×10^{-3} g/cm³, the mass of the air would be 1.2×10^{-3} g/cm³ × 2.1×10^{8} cm³ = 2.5×10^{5} g or 2.5×10^{2} kg. (This is equivalent to about 550 lbs.)

24 **Which of the substances listed in Table 3.2 are solids, which are liquids, and which are gases at (a) room temperature (20°C), (b) 50°C, (c) 100°C?**

This question is designed to make sure that your students understand Table 3.2. They should learn to study tables thoughtfully, not just give them a passing glance.
a) At 20°C, all the substances are liquids except *t*-butanol.
b) At 50°C, all the substances are liquids.
c) At 100°C, *s*-butanol has reached its boiling point; if heated at this temperature, it will boil. Cycloheptane and *n*-butanol are liquids. The remaining substances are gases.

25 **Figure B shows a diagram of a double boiler. Why is such a double boiler used to cook food that is easily scorched?**

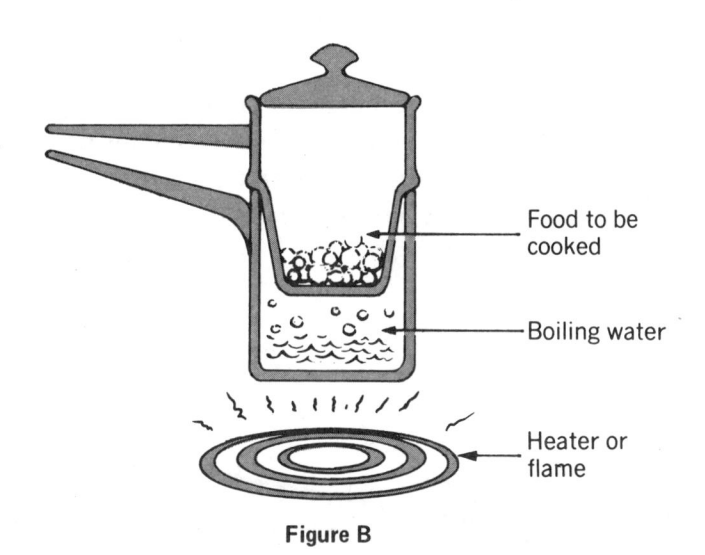

Figure B

The temperature of the water boiling in the lower section of the double boiler is constant (100°C at sea level). The food in the upper section of the double boiler will not be heated to a temperature higher than this.

26 **Object *A* has a mass of 500 g and a density of 5.0 g/cm³; object *B* has a mass of 650 g and a density of 6.5 g/cm³.**
a) **Which object would displace the most liquid?**
b) **Could object *A* and object *B* be made of the same substance?**

a) Volume of A = 500 g/(5.0 g/cm³) = 100 cm³
Volume of B = 650 g/(6.5 g/cm³) = 100 cm³
They will displace the same volume of liquid.
b) Their different densities show that they are not made of the same substance.

68 Characteristic Properties

27 A student measures the volume of a small aluminum ball by water displacement and then finds its mass on a balance. He finds that the sphere displaces 4.5 cm³ of water and has a mass of 6.5 g.
a) What value does the student obtain for the density of aluminum?
b) How might you account for the difference between this value for the density of aluminum and the one given in Table 3.1?

The main purpose of this problem is to have the student look again at Table 3.1.
a) Density = 6.5 g/4.5 cm³ = 1.4 g/cm³
b) Density of aluminum from Table 3.1 is 2.7 g/cm³. We are told that the ball is aluminum. If we assume that the mass and the volume were measured accurately enough so that the second digits in these measurements are significant, we can only conclude that the sphere is displacing too much water for its mass. The ball must be hollow.

28 A student has several specimens of substance A of different sizes and also of substance B of different sizes. She measures the masses and volumes of these specimens and plots the graphs shown in Fig. C. Which substance has the greater density? How do you know?

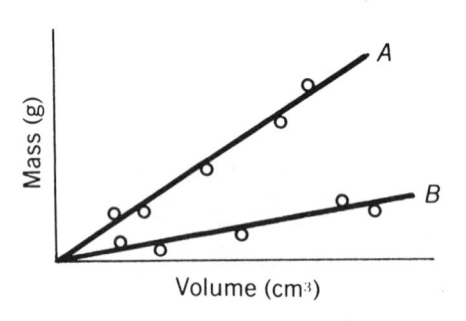

Figure C

Substance A has the greater density. You can see from the graphs, if you pick the same volume for both substances, that substance A has the greater mass. Since the density is found by dividing the mass by the volume, substance A has the greater density.

29 How would you determine the density of ice? Could you get the volume by melting the ice and measuring the volume of the resulting water?

The density of ice could be determined by finding the mass of the ice on a balance and then measuring its volume by the displacement of some fluid in which it will sink and in which it is not soluble, such as kerosine. Of course, this would have to be done at a temperature below the melting point of ice.

If we melted the ice and determined the volume of the substance this way, we would not be determining the density of ice; we would be determining the density of water, which is about 10 percent greater than that of ice close to its melting point.

30 **How would you distinguish between unlabeled pint cartons of milk and of cream without breaking the seals?**

The simplest way to distinguish between unlabeled pints of milk and cream would be to put the cartons on the opposite pans of an equal-arm balance large enough to take them. The carton containing the milk would have the greater mass.

Some students may be confused by the commercial labeling of cream as "heavy" and "light." The density of "heavy" cream is *less* than that of "light" cream, and both are less dense than milk or water. The "thickness" (viscosity) of the cream is not the same property as density! The exact meaning of these words is not the important point here; the operation (massing on the balance) needed to answer the question is the point. Comparing the mass is sufficient here because both containers have the same volume.

31 **Weight the end of a test tube with just enough sand so that it floats upright in water. With a pencil that marks on glass, mark the depth to which it sinks. How deep does it sink in alcohol? How could you use this device (called a "hydrometer") to measure the densities of unknown liquids?**

The tube will sink deeper in alcohol than in water. If the device is placed in several liquids of known density and the position of the liquid surface while it is floating at rest is marked each time on the tube, densities can be assigned to the marks and a scale constructed. The density of any liquid in which the device will float upright can then be read. This device is to be considered as a "black box"; that is, a device that can be used to perform an operation or make a measurement but whose principles of operation are not known. Digressions to discuss Archimedes' principle are not appropriate here. The graph in Fig. X shows the depth to which a test tube weighted with sand sinks in different liquids. The points were determined experimentally, and the curve was drawn by interpolation. As the graph shows, a density scale marked on the tube would not be linear. (The graph is not a straight line; equally spaced marks would not represent equal density difference.) Even without marking a scale on the test tube, students can use this crude hydrometer as a quick way to compare the densities of two liquids. If the test tube floats deeper in one liquid than in another, they know the liquids have different densities and are not the same substance.

When weighted with sand, the test tube will not float quite vertically, but it can still be used to compare the densities of two liquids. If it is weighted with lead shot so that it is one-half to two-thirds submerged in water, it will float vertically over the range of densities shown by the curve in the figure.

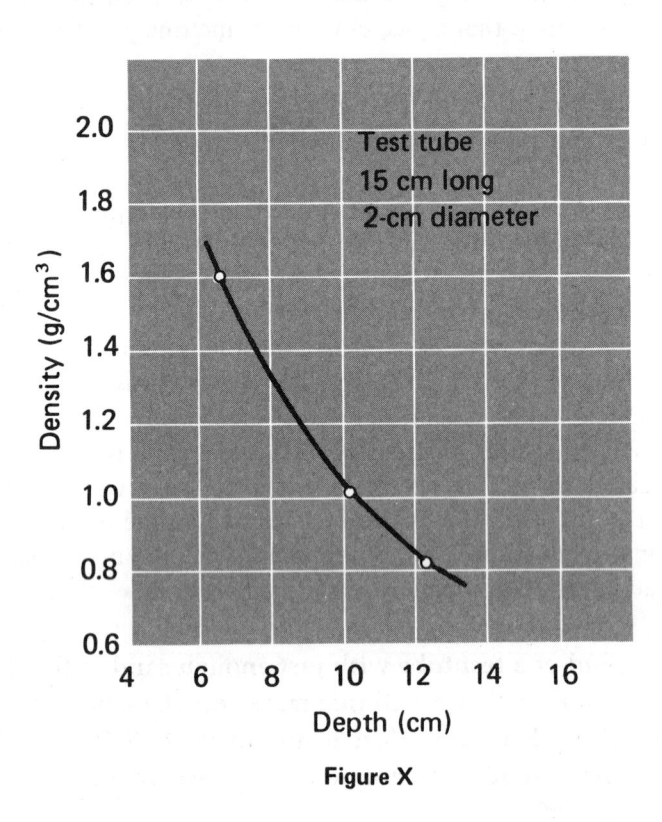

Figure X

Test tube 15 cm long 2-cm diameter

(Graph: Density (g/cm³) vs Depth (cm), with points at approximately (6, 1.6), (10, 1.0), (12, 0.8))

32 In Table 3.1, why are the pressure and temperature stated for the densities of gases and not for the densities of solids and liquids?

The volumes of gases change considerably with temperature and pressure, and so their densities depend on what the temperature and the pressure happen to be. On the other hand, the volumes of liquids and solids change only slightly with changes in temperature and pressure, and it usually is unnecessary to give the pressure and temperature when stating their densities.

33 The students in an IPS class in one of the coastal cities of the United States measured the boiling point of water and found that it was 100°C. On the same day the students in an IPS class in one of the mountain cities in the United States also measured the boiling point of water and found that it was 95°C. What can be inferred about the boiling point of water from these reports?

Evidently the temperature of boiling water depends on the altitude at which it is measured. Some or all the students may know that the pressure of the atmosphere decreases with increasing altitude; they may infer that the boiling point of water is affected by the atmospheric pressure, being lowered at reduced pressure.

34 A cylinder is closed with a tight-fitting piston 30 cm from the end wall (Fig. D); it contains a gas that has a density of 1.2 × 10⁻³ g/cm³. The piston is pushed in until it is 10 cm from the end wall; no gas escapes. What is the density of the compressed gas? What is your reasoning?

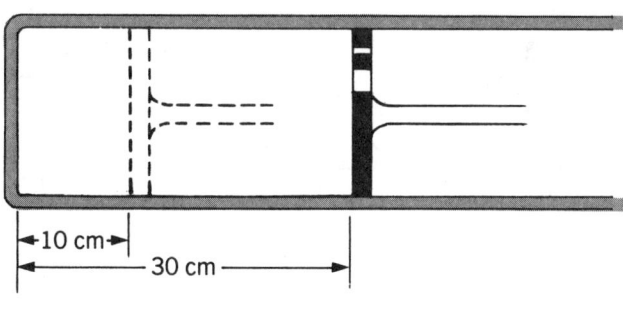

Figure D

Since the cross-section of the cylinder does not change, moving the piston from 30 cm to 10 cm from the end wall will reduce the volume to 1/3 of its original volume. Since no gas escapes, the mass of the gas is unchanged. Its density was M/V_i, (where V_i is the initial volume). Its new density is M/V_f, where $V_f = (1/3)V_i$ or $\frac{M}{(1/3)V_i} = 3\frac{M}{V_i}$, which is three times the original density. Therefore, the new density is 3 × 1.2 × 10⁻³ g/cm³ = 3.6 × 10⁻³ g/cm³.

35 Does the density of air change when it is heated:
a) In an open bottle?
b) In a tightly stoppered bottle?

a) When air is heated in an open bottle, it expands and some gas escapes, but the volume of gas in the bottle remains the same. Since the mass of the hot gas in the bottle is less, the density of the gas in the bottle will be less. You can easily demonstrate the expansion of the air with an apparatus like that shown in Fig. XI. The bottle also will expand, of course but much less than the gas, and this effect can be neglected for all practical purposes.

72 Characteristic Properties

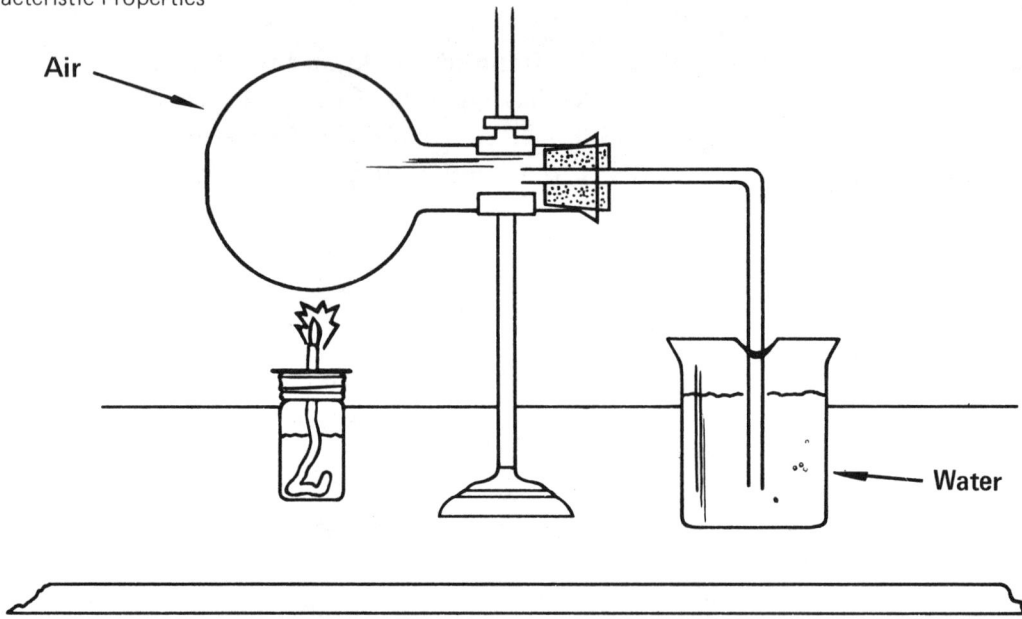

Figure XI

b) In a tightly stoppered bottle, the mass of the air would not change; the volume increases only very slightly from expansion of the bottle. Hence, the density would remain very nearly constant.

Solubility

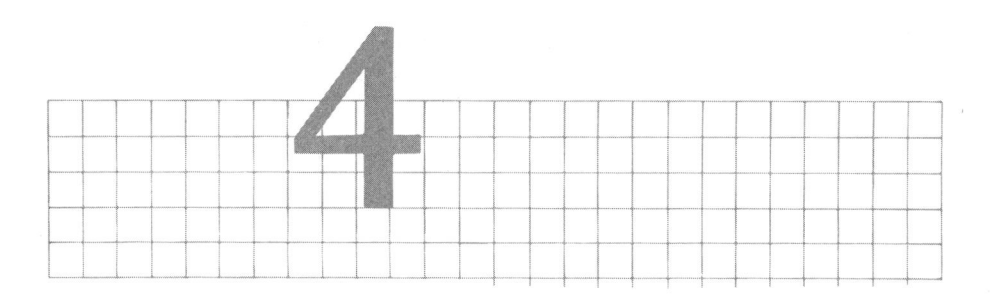

4

General Comments

Solubility is a characteristic property of both the solute and the solvent. It is expressed in a complex unit—grams of solute per 100 cm^3 of solvent. If we know the solubility of a substance in a given solvent and the quantity we want to dissolve, we can calculate the minimum amount of solvent necessary. Or, if we know how much solvent we have, we can use the solubility to find the maximum amount of the solute we can dissolve in it.

Like density, solubility changes with temperature, but whereas the density of solids or liquids changes only slightly with temperature, the solubility of some substances changes rather drastically. The dependence of solubility on temperature is very useful in separating substances in solution.

It will be helpful in the following chapter if your students understand and can use the graphs in Fig. 4.3. However, spend only a reasonable amount of time on the mechanics of solubility problems, and do not let yourself and your class get bogged down in them.

The suggested schedule for this chapter is:

Sections 1-2 (one expt., probs. 1-3; 34)	2 periods
Sections 3-4 (two expts., probs. 4-13; 24-28)	4 periods
Sections 5-6 (one expt., probs. 14-16; 29)	2 periods
Sections 7-10 (one expt., probs. 17-19; 30-32)	3 periods
Sections 11-13 (one expt., probs. 20-23; 33)	2 periods
Achievement Test No. 2	2 periods
	Total 15 periods

EXPERIMENT: DISSOLVING A SOLID IN WATER 4.1

This experiment introduces students to solutions with the use of potassium dichromate and water. The color of this solid, even in relatively dilute solutions, is sufficiently apparent so that the students can see the uniform distribution of the substance throughout the water. The students also can observe that there is a limit to the amount of solid a given quantity of water will dissolve.

74   Solubility

The Experiment

The solubility of potassium dichromate is temperature dependent, as seen in the following table.

Solubility of $K_2Cr_2O_7$ (g/100 cm³ of water)

Temperature (°C)	Solubility
15	9.8
20	12.2
25	15.0
30	18.0

Because there will probably be some variation in the final temperatures of the samples of water, you can expect some variation in the results reported by your students.

It would be best to dispense the water from containers that have been standing for some time in the room so that the water is near room temperature. Holding the test tubes while they are being shaken may result in warming the tubes to a temperature of around 25°C.

Provide each student group with about 6 g of potassium dichromate. The students should stopper the tubes and shake them vigorously until all the solid dissolves, or until it is evident that the remaining solid is not going to dissolve.

Sample Data

Total mass of orange solid	In 5.0 cm³ water	In 20.0 cm³ water
0.30 g	All dissolves	All dissolves
0.60 g	All dissolves	All dissolves
0.90 g	Solid remains	All dissolves

If the temperature of the water is much different from 25°C, you can use the solubility table to estimate what is likely to happen; however, do not yet discuss temperature effects with your students.

Answers to Questions

When 0.30 g of the orange solid is added to 5.0 cm³ and to 20.0 cm³ of water, all of the solid dissolves.

In each test tube the color is uniform. Since the color is uniform throughout each test tube, the mass dissolved in each cm³ of water in a given test tube is the same.

The shade of color is less intense in 20.0 cm³ of water than it is in 5.0 cm³ of water because in 20.0 cm³ of water there is less solid dissolved in each cm³ of water than there is in 5.0 cm³ of water.

Yes, 0.60 g of the orange solid dissolves completely in each tube.

The color is still uniform in each solution. The shade of color is less intense in 20.0 cm³ of water than it is in 5.0 cm³ of water.

After another 0.30 g of orange solid is added, there is then 0.90 g in each tube. After several minutes of shaking, there is undissolved solid in 5.0 cm³ of water. All the solid has dissolved in 20.0 cm³ of water.

We assume that each cm³ of water will dissolve the same mass of orange solid. To have some solid remain in 20.0 cm³ of water, we predict that we should add $\frac{20.0 \text{ cm}^3}{5.0 \text{ cm}^3} \times 0.9 \text{ g} = 3.6 \text{ g}$. When this amount was added, there was some solid left undissolved.

(Actually at 25°C, anything over 3.0 g of orange solid in 20.0 cm³ of water will not dissolve, but your students have no way of knowing this.)

Apparatus and Materials

2 Test tubes (20 × 150 mm)
　Graduated cylinder, 50 ml
　Balance
2 Paper massing cups
2 No. 2 solid stoppers

Test-tube rack
Scoopula
Water, room temperature
Potassium dichromate,
　　6 g per student group

CONCENTRATION　　4.2

We wish to arrive at the concept of solubility, a characteristic property of a particular combination of a substance and a liquid. As an intermediate step, it is convenient to introduce *concentration* as a way of expressing the quantity of substance dissolved in a fixed volume of the liquid. We can then use the idea of concentration in the following section to arrive at the concept of solubility.

EXPERIMENT: COMPARING THE CONCENTRATIONS OF　　4.3
SATURATED SOLUTIONS

The purpose of this experiment is to demonstrate another characteristic property of matter that can be used to distinguish between two substances. Sodium chloride and sodium nitrate appear similar in many respects: in granular form they are both white, and they have about the same density. The fact that they are different substances is not immediately apparent, but by finding the concentration of saturated solutions of each, we see that there is a significant difference in the amounts that will dissolve in a given volume of water. The concentration of a saturated solution of a substance is called its solubility.

The Experiment

Furnish each student group with two solids, labeled *A* and *B*, and tell each group which solid they are to use in the quantitative portion of the experiment. Use sodium chloride for solid *A*, and preferably sodium nitrate for solid *B*, but sodium chlorate may be substituted for the nitrate. If sodium nitrate is used for solid *B*, have your students use about 5 g of each solid; if sodium chlorate is used, have your students use about 6 g of each solid. It is recommended that the solids be of reagent grade. If a lower grade is used, a cloudy solution may result. When the water (5 cm³)

76   Solubility

is added and the tubes are shaken, it is easy to see that more of one substance has dissolved than of the other.

Your students should shake the test tubes of solid and water vigorously for at least 5 minutes. The mixtures cool during the dissolving process. When they have warmed again to room temperature, the saturated solutions can be decanted into the massed evaporating dishes. The evaporation to dryness will involve some spattering, but most of this can be avoided if your students heat the solutions slowly.

Sample Data

Solution temperature: 26°C	Sodium Chloride	Sodium Nitrate	Sodium Chlorate
Mass of solution (g)	4.96	6.08	8.99
Mass of remaining solid (g)	1.26	2.98	4.73
Mass of water (g)	3.70	3.10	4.26
Mass of solid (g)/mass of water (g)	0.34	0.96	1.11
Mass of solid (g)/100 cm^3 water	34	96	111

Answers to Questions

One sample of solid is obviously more soluble in water than the other. Therefore, the samples are of different substances.

Apparatus and Materials

Balance and massing cups
2 Test tubes (20 X 150 mm)
Scoopula
Stirring rod
Graduated cylinder (10 ml)
Burner and stand
 Sodium nitrate or
 Sodium chlorate (5–6 g)

Sodium chloride (5–6 g)
2 No. 2 solid stoppers
Test-tube rack
2 Small evaporating dishes
Safety glasses

4.4   EXPERIMENT: THE EFFECT OF TEMPERATURE ON SOLUBILITY

The two solids used in Expt. 4.3 are obviously different substances, since they have different solubilities in water at room temperature. But it is possible to find two solids that have about the same solubilities at this temperature. This does not mean, though, that they will have the same solubility at other temperatures. We may, therefore, distinguish between them by determining their solubilities at a different temperature. The fact that the solubilities of two substances change differently with temperature will also make it possible to separate one substance from the other. The ideas in this section are important for an understanding of Expt. 5.5, The Separation of a Mixture of Soluble Solids, in Chap. 5.

An interested class can actually perform the potassium sulfate experiment and construct Fig. 4.2 by having a different temperature assigned to each group in the class.

The Experiment

Considerable stirring is necessary to make sure that the solutions are saturated at room temperature. Both test tubes will contain considerable amounts of undissolved salt at first. When heated to the boiling point of water, all the potassium nitrate will dissolve, but there will be no apparent change in the amount of sodium chloride dissolved.

Answers to Questions

The solubilities of the two solids appear to be quite different as the temperature is changed from room temperature to the boiling point. When the test tubes are allowed to cool in a beaker of cold water, a considerable amount of one of the solids precipitates out of solution.

Apparatus and Materials

Sodium chloride (10 g)
Potassium nitrate (10 g)
2 Test tubes (20 × 150 mm)
Test-tube rack
Beaker (250 ml)
Graduated cylinder (10 ml)
Burner and stand
2 No. 2 solid stoppers
Safety glasses

WOOD ALCOHOL AND GRAIN ALCOHOL 4.5

So far we have limited the discussion of solubility to water, the most common solvent. However, there are a host of other solvents available to assist in identifying substances and in separating one substance from another. Here we discuss two of the solvents known from ancient times—methanol and ethanol.

You might point out to your students that we could perform quantitative experiments like those in Sections 4.3 and 4.4 with these solvents as well; however, we do not need to repeat such experiments, since procedures and results would be analogous to those for water. Data on these and other solvents are available in chemical handbooks and manuals.

EXPERIMENT: METHANOL AS A SOLVENT 4.6

In Expt. 4.3 students found out that the solubility of substances in water is a characteristic property. Now they find out that the solubility of substances in another liquid (alcohol) is also a characteristic property. The use of different solvents increases considerably the usefulness of solubility for distinguishing between substances. Of course, the procedure can be reversed. For example, we can distinguish between methanol and water by the solubility of sugar in each.

The Experiment

It is possible to complete this experiment in one class period. The instructions have been left open to encourage students to design their own experiment. You may wish to ask students to prepare a plan of attack for your approval before proceeding with the experiment. On the other hand, you may wish to develop a "class plan" through a class discussion in the pre-lab. Regardless of the approach, your students should be able to organize the activities and defend their approach to the experiment.

Tell your students which solid is which before they investigate their solubilities. There are five solids to be dispensed. This saves considerable confusion in the laboratory. In this experiment it is necessary only to compare solubilities. It is not necessary to get quantitative results. The best way to compare quickly the solubilities of citric acid and sugar in methanol is to set up two tubes containing 10 cm^3 each of methanol. Burner fuel can be used if methanol is not available. Then add a pinch of sugar (a small amount on the end of a spatula) to one test tube of methanol and a pinch of citric acid to the other. (All "pinches" should be about the same size.) Considerable shaking may be necessary to dissolve the substance. If this process is continued, shaking the stoppered test tube after each pinch of substance is added, it can easily be seen that sugar is not very soluble in this liquid and that citric acid is quite soluble. Do not tell your students what will happen. Just be sure that they add only a small amount of each substance at a time.

In testing the solubility of moth flakes, magnesium, and magnesium carbonate in water and in methanol, the same procedure is used.

As a supplement to this experiment, or as a short laboratory test, you may wish to have the students try to identify three unknowns. You can give them salt, sugar, and citric acid, labeled only as x, y, and z.

Answers to Questions

Sugar and citric acid can be distinguished easily by their solubilities in methanol, because sugar is only slightly soluble in this liquid while citric acid is very soluble.

Moth flakes appear to be insoluble in water but soluble in methanol.

Magnesium and magnesium carbonate are insoluble in both water and methanol.

(Magnesium carbonate forms a suspension of undissolved solid particles in water, which at first will appear cloudy but will settle after standing. Your students may ask whether the substance has dissolved. A cloudy solution indicates that some material is undissolved. For the purposes of this experiment, if there is less solid after shaking than before, then some solid has dissolved. Students can only conclude that little or no magnesium carbonate dissolves in water.)

The results are summarized in the following table.

Substance	Solubility in Water	Solubility in Methanol
Sugar	Very soluble	Slightly soluble
Citric acid	Very soluble	Very soluble
Moth flakes	Insoluble	Soluble
Magnesium	Insoluble	Insoluble
Magnesium carbonate	Insoluble	Insoluble

Apparatus and Materials

4 Test tubes
  Test-tube rack
  Spatula
  Graduated cylinder (10 ml)
2 No. 2 solid stoppers
  Sugar (about 3 g)
  Citric acid (about 4 g)

Moth flakes (about 3 g)
Magnesium ribbon (about 1 cm)
Magnesium carbonate (about 2 g)
Methanol or burner fuel (50 cm^3)
Water
Paper towels

SULFURIC ACID 4.7

Sulfuric acid was originally called oil of vitriol because of its method of preparation and its properties. Early experimenters knew only that by heating certain minerals, a liquid was produced which had useful properties. Substances from which this "oil" could be obtained were named *vitriol* and distinguished only by color. The green vitriol in this experiment is ferrous sulfate. The historical name is used on purpose to emphasize that sulfuric acid was successfully used long before its chemical composition was known. What counts is what it does and not what it is called. It is worthwhile to emphasize this point in class. You may wish to prepare sulfuric acid from green vitriol (see Fig. I) as a class demonstration.

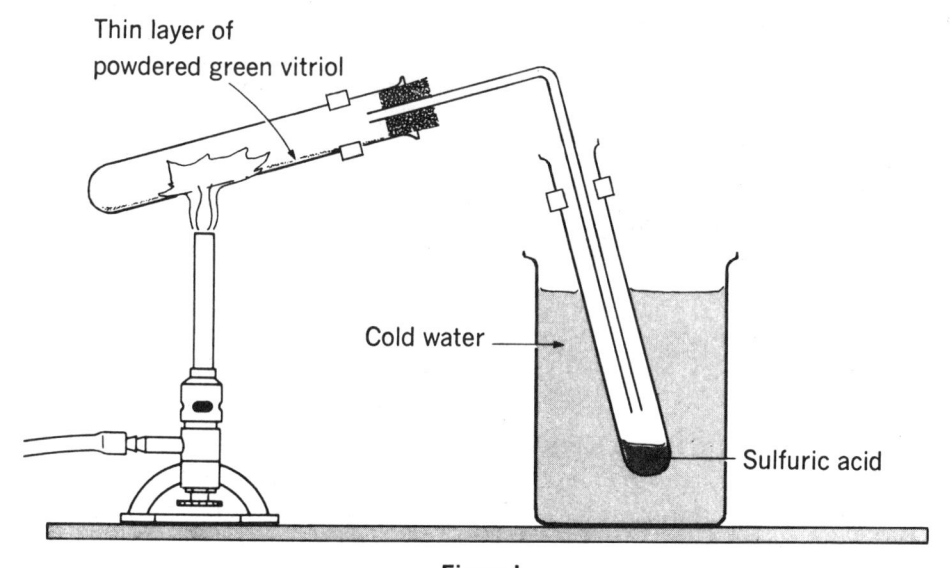

Figure I

Demonstration

About a fourth of a test tube (7 g) of green vitriol is all that can be heated sufficiently to yield acid. It must be finely powdered and spread as thinly as possible over nearly the whole length of a Pyrex test tube so that all of it will be strongly heated. Strong heating is very important in this demonstration. It is recommended that Bunsen burners (or propane torches) be used. Two alcohol burners can be used, but the yield of acid may be very small.

The burner flame should be moved to a new spot under the test tube as soon as the vitriol over it has become dark brown. When the vitriol is heated, it first changes color from green to white as it loses its water of hydration, which condenses in the test tube soon after the heating is started. Stronger heating then produces a whitish vapor (sulfur trioxide) which dissolves in the water already condensed in the collecting tube to produce sulfuric acid. The vitriol turns brown in decomposing into sulfur trioxide and iron oxide. The appearance of the brown oxide and the whitish vapor is evidence that the vitriol is sufficiently hot to produce sulfuric acid. If the heating is not sufficiently strong, this will not happen, and only water will be collected.

The yield of sulfur trioxide is small compared with the amount of water collected, and so the solution is very dilute. For this reason only tiny amounts of magnesium and magnesium carbonate should be placed in the acid. Large amounts will not show any appreciable change in amount of solid before all the acid is used up in the reaction. You should use a magnified projection of the test tube of acid on a screen or the chalkboard. The pieces of magnesium and magnesium carbonate are so small that otherwise it is difficult to see them dissolve. The projection can easily be done by placing a watch glass containing the acid and substance to be dissolved on an overhead projector.

Some of your colleagues may object to the use of the word "dissolved" to describe what happens when we put a piece of magnesium in sulfuric acid. For us, and for your students in the course, the disappearance of a solid in a liquid means that the solid has dissolved. Further study shows that in some cases we can recover the substance simply by evaporating the liquid, whereas in others we end up with a different substance. We often distinguish the latter cases from the former by saying that a "chemical reaction" takes place, but such a distinction is not useful here. Moreover, it is not quite correct: Even when the original substance can be recovered, chemical reactions may take place in the process of dissolving.

Apparatus and Materials

Pegboard
Bunsen burner or propane torch
2 Test tubes
Watch glass
Test-tube rack
2 Clamps
Beaker
No. 2 one-hole stopper
Glass bend

Magnesium ribbon (3 mm)
Magnesium carbonate
    (about a 3-mm cube)
Green vitriol (hydrated ferrous
    sulfate, $FeSO_4 \cdot 7H_2O$), 7 g
Water
Paper towel
Safety glasses
Overhead projector

4.8 EXPERIMENT: TWO GASES

The two gases produced in this experiment are investigated, using flammability and the effect of the gases on limewater. These characteristic properties have not been discussed before, but the students should have no difficulty in recognizing them as such.

Again the emphasis is on the properties of unfamiliar substances that show that they are different. The names of the gases are given only after the experiment has been completed.

The gases are collected in test tubes over water from the reaction of magnesium and magnesium carbonate with 1N sulfuric acid.

The density of the gases is not measured quantitatively, but is compared qualitatively with the density of the air. If the gas disappears from a test tube held upright, its density is less than that of air. If it disappears from an inverted test tube, it is more dense than air.

To make the density test conclusive, open test tubes of gas should be held in the test position for about 60 seconds before testing the gas with a glowing or burning splint.

The Experiment

It is necessary to prepare 1N acid before the class period by dissolving 20 cm^3 of concentrated sulfuric acid in 700 cm^3 of water.

CAUTION: A large amount of heat is evolved in dissolving the acid; therefore, the acid should always be added to the water a little at a time, with constant stirring. Do *not* add the water to the acid, since in this case the water will boil locally and spatter concentrated acid around.

Limewater should be prepared several days before it is used. Add 2 g of calcium hydroxide to 1,000 cm^3 of water. Stir thoroughly, and set the solution aside for at least a day. The clear limewater can then be carefully decanted into a bottle. Keep the bottle stoppered; otherwise, carbon dioxide in the air will slowly react with the limewater and produce a thin scum of calcium carbonate on the surface.

Four test tubes of each gas will be needed to make the flammability tests. Instruct your students to arrange two small test tubes of water inverted in the plastic bucket of water and held just under the surface by two small clamps. The tubing must be switched quickly from one test tube to another. If you have enough glass bends, it will help in the collection of gas to insert a glass bend in the end of the tubing from which the gas is expelled.

The magnesium and sulfuric acid are reacted in a large test tube. Roll up the magnesium-ribbon strips, and place them in the test tube first. The stopper and tubing should be in place ready to collect gas before the sulfuric acid solution is added. Add about three-fourths of a test tube of acid solution (about 25 cm^3), pouring it quickly and stoppering the test tube immediately. While one student is collecting gas, the other can stopper and remove full tubes of gas and replace tubes of gas with tubes of water until four tubes of gas are collected.

The magnesium carbonate and sulfuric acid are also reacted in a large test tube. Break magnesium carbonate blocks into chunks that are just large enough to fit into the test tube. *Do not use powdered magnesium carbonate*, because it generates gas very rapidly. Put the magnesium carbonate into the test tube first. Pour the acid solution (about 35 cm^3) into the tube quickly and stopper immediately. If the gas is collected efficiently, more than four test tubes of carbon dioxide will be collected.

Both glowing and burning splints are used to make the flammability test conclusive. A burning splint placed in a test tube of air uses up the oxygen very rapidly and goes out. Therefore, it is not easy to distinguish between what happens with a burning splint in a test tube of air and what happens in a test tube of carbon dioxide.

The density tests will be much more reliable if the open test tubes are held in the upright or upside-down position for 60 seconds before making the splint tests. It will save time and afford quick comparison if you hold two test tubes of gas together, one upright and one inverted, for the 60-second period, and then apply the splint test to each in turn.

Although students are not instructed to make the flammability test with air, some of them will probably wonder if perhaps one of the gases they have produced shows the same lack of flammability as air. If this is the case, they should, of course, try the test with air.

When carbon dioxide is bubbled into limewater, the water at first becomes cloudy with calcium carbonate. If enough carbon dioxide is bubbled through the water, it becomes clear again as a result of the insoluble carbonate changing to the soluble bicarbonate. This may happen if students bubble the gas through the limewater for a long time, but it in no way invalidates the limewater test for carbon dioxide.

Answers to Questions

The gas collected in the first test tube should be discarded because it probably contains a large amount of the air originally present in the test tube before the gas was generated.

An inverted test tube of the gas produced by magnesium burns with a "pop" when a flaming splint is inserted. A test tube of the gas that has remained right-side up for one minute does not give any "pop" when a flaming splint is inserted.

When a glowing splint is used with the gas produced by magnesium in a test tube that is right-side up, nothing happens. In an inverted test tube, the glowing splint may go out, or the gas may ignite with a "pop." This depends on how much air is mixed with the hydrogen. (A flaming splint always produces convection currents that mix enough air with the hydrogen to cause rapid burning, whereas a relatively cool glowing splint may not.)

The gas produced by magnesium is less dense than air, since it rises out of the upright test tube and is replaced by air, which does not burn.

When the gas produced by the magnesium is bubbled through limewater, nothing happens.

If the experiment is repeated using magnesium carbonate and sulfuric acid, both a glowing splint and a flaming splint go out very quickly when inserted into upright test tubes of the gas. A glowing splint continues to glow when inserted in an inverted test tube that contained the gas, while a flaming splint soon goes out (as it will in a test tube of air). The gas is more dense than air, since it flows out of an inverted test tube.

When bubbled through limewater, the gas produced by the magnesium carbonate turns the limewater milky.

(If your students have trouble distinguishing the gases from each other on the basis of the splint tests, a demonstration of the tests for each gas in the post-lab discussion is helpful.)

The results of the experiment are shown in the following table.

Substances	Flammability				Limewater test
	Mouth Up (60 sec)		Mouth Down (60 sec)		
	Glowing splint	Flaming splint	Glowing splint	Flaming splint	
Magnesium and sulfuric acid	Glows	Finally goes out	Goes out or gas ignites with a "pop"	Gas burns with "pop"	Clear
Magnesium carbonate and sulfuric acid	Goes out	Goes out	Glows	Finally goes out	Cloudy
Air	Glows	Finally goes out	Glows	Finally goes out	Clear

Apparatus and Materials

Pegboard
2 Small clamps
6 Small test tubes
2 Large test tubes
  Test-tube rack
  Plastic bucket or large pan
  No. 2 one-hole stopper
  No. 4 one-hole stopper
4 No. 2 solid stoppers
2 Right-angle glass bends
  Rubber tubing
  Dilute sulfuric acid
    1N (60 cm³)

Limewater (10 cm³)
Magnesium ribbon
  (five 7-cm strips)
Magnesium carbonate
  (2 g in several pieces)
Matches
Wood splints
Water
Paper towel
Safety glasses

HYDROGEN 4.9
CARBON DIOXIDE 4.10

These two sections are designed to summarize in a quantitative manner the characteristic properties of the gases produced in Expt. 4.8 and to show once more, in the case of hydrogen, how substances get their names. Put the emphasis on the difference in their characteristic properties and not on the memorization of the numerical values of the densities, boiling points, etc.

84   Solubility

4.11 THE SOLUBILITY OF GASES
4.12 EXPERIMENT: THE SOLUBILITY OF AMMONIA GAS

The gases studied at first were either collected over water or prepared from an aqueous solution. This fact implies that the gases are not seriously affected by the presence of water. However, this is not true of all gases, as this particular experiment is designed to show. Solubility is a characteristic property that can be used to distinguish between different gases, just as between different solids.

As with solids, some gases are much less soluble than others, and it would be very difficult to determine the extremely small amount of nearly insoluble gases that actually dissolves. These gases can be prepared or collected over water. On the other hand, gases that are very soluble in water must be kept away from water if we are to collect them.

Unlike the situation with most solids, the solubility of the gas decreases when the temperature of the solution is raised. In such cases we usually are able to obtain some of a soluble gas from a solution simply by heating.

The Experiment

When your students have set up the apparatus shown in Fig. 4.5, make sure that the upper test tube is dry. As the solution is heated with the alcohol burner, the dissolved ammonia gas will begin to bubble out. The heat should be adjusted so that the solution continues to bubble very gently. If the solution is heated too vigorously, nearly all the ammonia gas will be driven off; then the gas that escapes from the solution becomes mostly water vapor, which condenses in the inverted test tube and dissolves most of the ammonia gas originally collected. Water vapor condensed on the inside of the collecting tube will also dissolve the ammonia gas as fast as it is liberated from the solution. The upper test tube will be quite full if the gas is collected for about one minute after the ammonia solution starts to bubble. Remove the tube slowly so that the ammonia gas is not disturbed.

Commercially available ammonia solution can be used at full strength, or a solution can be prepared by diluting concentrated ammonium hydroxide with two parts of water.

If the test tube is closed by a rubber stopper and placed in a beaker of water, as directed in the text, the stopper must not be too tight to be removed in one quick movement. Any delay will allow a small amount of water to seep into the test tube before the stopper is all the way out. The ammonia gas will immediately dissolve in this water and create a partial vacuum that sucks the stopper back firmly into the tube. It is then very difficult to dislodge the stopper.

This can be avoided by having your students loosely cover the test tube with a large stopper, a glass plate, or a piece of heavy paper. In this case it will be easier to place the test tube in a bucket than in a beaker. Shaking the tube assures complete reaction between the ammonia and the water. Any air that remained in the test tube will form a bubble at the top of the tube.

Students may want to try to collect the gas by the displacement of water. At first, air bubbles into the collecting tube; as soon as some ammonia is generated, the gas dissolves and water backs up into the generating tube. No harm is done, and the solubility of ammonia is clearly demonstrated.

The "ammonia fountain" is a classic demonstration of the solubility of ammonia in water. Fill a dry 500-ml Florence flask with ammonia gas by the same technique your students used to fill their test tubes. Insert a stopper fitted with a thin glass tube, and invert the flask so that the end of the tube is under water in a beaker as shown in Fig. II. Water rises in the tube and sprays like a fountain into the flask.

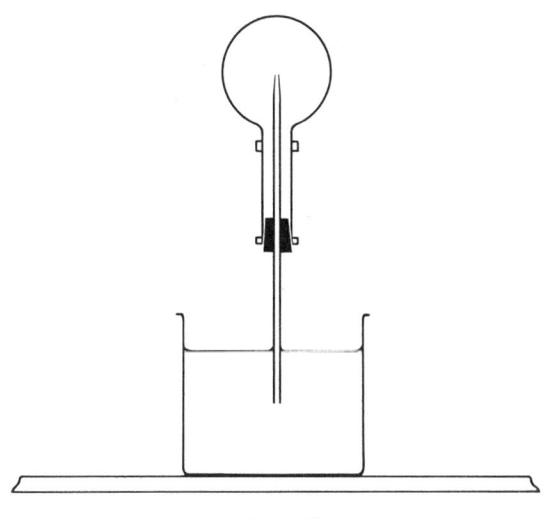

Figure II

In the post-lab you may wish to compare the solubilities of various gases. The following table indicates the solubility of gases in water at 20°C and at atmospheric pressure.

Gas	$\dfrac{cm^3 \text{ gas}}{cm^3 \text{ water}}$	$\dfrac{g \text{ gas}}{100 \text{ cm}^3 \text{ water}}$
Hydrogen	0.018	0.00016
Oxygen	0.031	0.0043
Nitrogen	0.015	0.0019
Carbon dioxide	0.878	0.169
Ammonia	715	54
Chlorine	1.77	0.57
Sulfur dioxide	39.37	11.28

Sample Data

Trial	Volume of air bubble remaining (cm³)	Percent of test tube filled with water
No. 1	3.0	91
No. 2	2.6	93
No. 3	4.0	89

86 Solubility

Answers to Questions

Since ammonia gas is obtained from a water solution simply by heating, we can conclude that the solubility of the gas decreases as the temperature of the solution is raised. The ammonia gas is successfully collected in an inverted test tube, which indicates that it is less dense than air. The ammonia gas rises in the test tube, displacing the air, which is forced out the bottom.

Apparatus and Materials

Pegboard	No. 2 one-hole rubber stopper
2 Small clamps	No. 2 solid rubber stopper,
2 Small test tubes	or glass plate
Bucket (or beaker)	Straight piece glass tubing
Alcohol burner	Water
Burner stand	Matches
Boiling chips	Paper towel
Ammonia solution (15 cm^3)	Safety glasses

4.13 OTHER SOLVENTS

You may wish to demonstrate the action of hydrochloric and nitric acids on metals. Use copper, since it is mentioned in the text. You might also try zinc and aluminum. Avoid using granular or powdered metals, since the reactions may be too rapid to be controlled.

CHAPTER 4—ANSWERS TO PROBLEMS

Sec.	Easy	Medium	Hard	Class Discussion	Home or Lab
1-2		1, 2, 3, 25			
3	6	4, 5, 24			
4	7, 10†, 12†	8, 9, 11, 13, 27, 28	26, 34	34	
5		14			
6	16	15†			29
7-10	19, 31	17, 18, 32			30, 31
11	20†	21, 22, 33			
12	23†				
13					

1. For Expt. 4.1, calculate the concentration of the solutions in g/cm³ and in g/100 cm³ after the addition of each sample of solid.

	Total mass of solid added to each tube	Concentration in 5.0 cm³ of water		Concentration in 20.0 cm³ of water	
Sample 1	0.30 g	0.060 g/cm³	6.0 g/100 cm³	0.015 g/cm³	1.5 g/100 cm³
Sample 2	0.60 g	0.12	12	0.030	3.0
Sample 3	0.90 g	Uncertain; not all dissolved		0.045	4.5

2. What can you say about the largest concentration you were able to make in the experiment?

   The largest concentration was about 12 g/100 cm³ of water.

3. A mass of 25.0 g of sugar is dissolved in 150 cm³ of water. What is the concentration in g/100 cm³?

   The concentration is $\frac{25.0 \text{ g}}{150 \text{ cm}^3} \times 100 = 16.7$ g/100 cm³ of water.

4. Mario wishes to construct a table that lists the solubility in water of various substances. From various sources he finds the following data for solubilities at 0°C.
   - a) Boric acid — 0.20 g in 10 cm³ of water
   - b) Bromine — 25 g in 600 cm³ of water
   - c) Washing soda — 220 g in 1,000 cm³ of water
   - d) Baking soda — 24 g in 350 cm³ of water

What is the solubility of each substance in grams per 100 cm³ of water?

a) 0.20 × 100/10 = <u>2.0</u>
b) 25 × 100/600 = <u>4.2</u>
c) 220 × 100/1,000 = <u>22.0</u>
d) 24 × 100/350 = <u>6.9</u>

5 From your answers to Question 4, find the largest mass of each substance that will dissolve in 60 cm³ of water.

a) Boric acid   2.0 × 60/100 = 1.2 g
b) Bromine   4.2 × 60/100 = 2.5 g
c) Washing soda   22.0 × 60/100 = 13.2 g
d) Baking soda   6.9 × 60/100 = 4.1 g

6 Suppose that 200 cm³ of a saturated solution of potassium nitrate were left standing in an open beaker on your laboratory desk for three weeks. During this time most of the water evaporated. Would the mass of potassium nitrate dissolved in the solution change? Would the concentration of the potassium nitrate solution change during the three weeks?

The mass of dissolved potassium nitrate would decrease. The concentration of the solution would not change (provided the temperature remains constant).

7 If you plotted the data from Fig. 4.2 on Fig. 4.3, where would the graph be found?

The graph would be found below the graphs of both potassium nitrate and sodium chloride. It would run from 9 g/100 cm³ at 10°C to 23 g/100 cm³ at 100°C.

8 Suppose you have a saturated solution of potassium nitrate at room temperature. From Fig. 4.3 what do you predict will happen if you
a) heat the solution?
b) cool the solution?

a) No change will occur if the solution is heated; however, it could dissolve more potassium nitrate.
b) Some solid potassium nitrate will precipitate out of the solution if the solution is cooled.

9 What reason can you give for heating the solutions in Expt. 4.4 by immersing the test tubes in a beaker of hot water rather than heating the test tube directly over a burner flame?

Solubility 89

The test tubes are heated together in the water bath so that both tubes will be at the same temperature. This makes possible a direct comparison of the effects of raising the temperature on the solubilities of the two solids.

10† What temperature is required to dissolve 110 g of sodium nitrate in 100 cm³ of water?

From Fig. 4.3: 47°C.

11 a) If 20 g of sodium chloride is dissolved in 100 cm³ of water at 20°C, is the solution saturated?
b) How do you know if a solution is saturated?

a) As can be seen from Fig. 4.3, the solution is not saturated.
b) No more of the solid will dissolve in the solution.

12† A mass of 30 g of potassium nitrate is dissolved in 100 cm³ of water at 20°C. The solution is heated to 100°C. How many more grams of potassium nitrate must be added to saturate the solution?

From Fig. 4.3: 212 g.

13 A mass of 10 g of sodium nitrate is dissolved in 10 cm³ of water at 80°C. As the solution is cooled, at what temperature should a precipitate first appear?

Ten grams of sodium nitrate dissolved in 10 cm³ of water is equivalent to 100 g dissolved in 100 cm³ of water. Figure 4.3 shows that if the solution is cooled to 37°C, it will be saturated, and so a precipitate will begin to form at this temperature.

14 You distilled wood in Expt. 1.1. Where would you expect to find the methanol at the end of the stage of the experiment shown in Fig. 1.2? In Fig. 1.3?

At the end of the stage of the experiment shown in Fig. 1.2, you would expect to find the methanol in the test tube immersed in water. The boiling point of methanol is about 65°C, and so it will turn into a vapor in the test tube that is heated and condense in the test tube that is in the water, where the temperature is about room temperature (25°C).

At the end of the stage of the experiment shown in Fig. 1.3, once again you would expect to find the methanol in the test tube immersed in water, and for the same reasons.

15† Two solids appear to be the same and are both insoluble in methanol. A student, whose results are reliable to 5 percent, reports the solubilities of the two solids in water, as in the following table. Are these solids the same substance? Explain your answer.

Solid	Solubility (g/100 cm³)	
	0°C	100°C
A	73	180
B	76	230

Although the solubilities at 0°C are within 5 percent of each other, the solubilities at 100°C differ by about 25 percent. The solids are not the same.

16 a) Which of the following substances, x, y, and z, do you think are the same?
b) How might you test them further to make sure?

Substance	Density (g/cm³)	Melting point (°C)	Boiling point (°C)	Solubility in water at 20°C (g/100 cm³)	Solubility in methanol at 20°C
x	1.63	80	327	20	insoluble
y	1.63	81	326	19	insoluble
z	1.62	60	310	156	insoluble

a) Substance z is clearly different from x or y, but substances x and y are almost certainly the same if the last digit is significant but not exact (as is usually the case).
b) To make certain, we might try dissolving x and y in other solvents, such as carbon tetrachloride or ether. (Of course, such solvents are dangerous for students to use.)

17 A student might try to find the density of hydrogen by the method you used for finding the density of a gas in Expt. 3.9, where you collected about 400 cm³ of gas.
a) What would be the mass of this volume of hydrogen?
b) What difference in mass would there be between the total mass of test tube, acid, and magnesium at the start and the total mass of test tube and contents at the end of the action?
c) Could you accurately measure this difference in mass on your balance?

a) Using the density of hydrogen given in Table 3.1, the mass of 400 cm³ of hydrogen is 8.4×10^{-5} g/cm³ $\times$ 400 cm³ = $\underline{3.4 \times 10^{-2}}$ g.

b) 3.4×10^{-2} g.
c) Yes, an error of about 15 percent in this mass might be expected if only the rider was moved to measure this change.

18 Describe how you would show that carbon dioxide is produced in a candle flame.

Have two jars with covers ready; in one, place some limewater and close the cover. In the other, place some limewater and a burning candle and close the cover. When the candle goes out, shake both jars to mix the limewater with the gases in the jars. You would be sure the burning candle produced carbon dioxide if the limewater turns milkier in the bottle containing the candle than it did in the bottle with air only.

19 Carbon dioxide is contained in many common fire extinguishers. Give two reasons why it is a good fire extinguisher.

Carbon dioxide neither burns nor supports the burning of other common flammable materials. It is heavier than air, and so it will exclude air from the fire by settling at the base of the fire.

20† In certain shallow parts of Long Island Sound fish have been found dead of oxygen starvation on extremely hot days, though at other times fish live there very happily. What property of oxygen do these facts suggest?

The solubility of oxygen in water decreases as the temperature of the water rises.

21 Experiment 3.9 was performed twice, but with one tablet and 15 cm^3 of water in each case. When the rubber tube was placed as shown in Fig. 3.6, 435 cm^3 of gas was collected. When the tube reached only slightly beyond the mouth of the bottle, only 370 cm^3 of gas was collected. The change in mass of the reactants was the same in both cases.
a) What volume of gas dissolved in the water?
b) Use Table 3.1 (page 43) to find what mass of gas dissolved in the water.

a) The volume of the gas dissolved in the water was 435 cm^3 − 370 cm^3 = 65 cm^3.
b) Carbon dioxide has a density of 1.8×10^{-3} g/cm^3 at atmospheric pressure and room temperature; thus the mass of dissolved gas is 1.8×10^{-3} g/cm$^3 \times 65$ cm^3 = 0.12 g.

22 The following table shows the solubility of carbon dioxide at atmospheric pressure at various temperatures.

Solubility (g/100 cm³)	Temperature (°C)
0.34	0
0.24	10
0.18	20
0.14	30
0.12	40
0.10	50
0.086	60

a) Draw a graph from these data.
b) Find, from your graph, how much carbon dioxide will dissolve in 100 cm³ of water at 25°C.
c) Use Table 3.1, page 43, to find the volume of carbon dioxide that will dissolve in 100 cm³ of water at room temperature (20°C).
d) How much carbon dioxide will dissolve in 100 cm³ of water at 95°C?

a) See the graph in Fig. III.

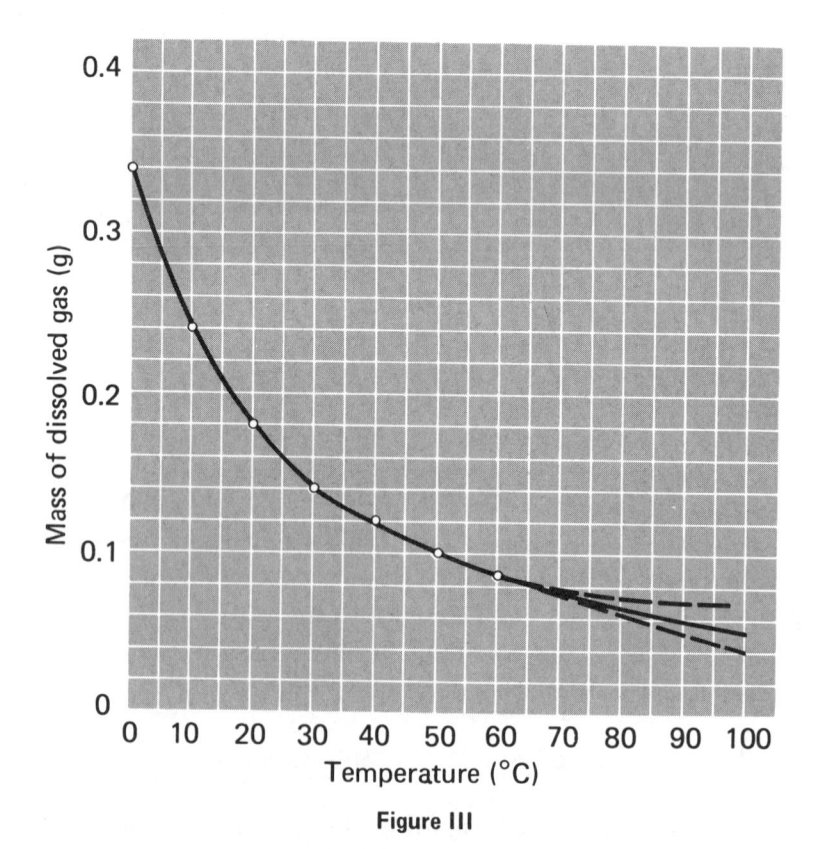

Figure III

b) At 25°C, about 0.160 g will dissolve.

c) The volume = $\frac{\text{Mass}}{\text{Density}} = \frac{0.18 \text{ g}}{1.8 \times 10^{-3} \text{ g/cm}^3} = \underline{100 \text{ cm}^3}$

d) The extrapolation of the graph will undoubtedly vary from student to student; there can be no one "correct" answer from the data given. The mass of carbon dioxide that will dissolve at 95°C probably lies between 0.04 g and 0.07 g.

23† If the tube of ammonia gas had been inverted and placed in water saturated with ammonia, would the liquid have risen in the test tube as it did in this experiment?

No. Water saturated with ammonia gas would not rise up into the tube because none of the ammonia gas would dissolve in water already saturated with the gas.

24 The solubility of chalk in water is 10^{-3} g/100 cm^3. How much water would be necessary to dissolve a piece whose mass is 5 g?

If your students understand the addition and subtraction of negative exponents, you can teach them at this point how to perform such calculations. If your students cannot learn this mathematical technique easily, have them convert numbers to standard notation and proceed accordingly, changing their final answers to powers-of-10 notation. Volume of water = (5 g/10^{-3} g) × 100 cm^3 = $\underline{5 \times 10^5 \text{ cm}^3}$.
    (In standard notation: (5 g/0.001 g) × 100 cm^3 = 5,000,000 cm^3 = 5 × 10^5 cm^3.)

25 Each of four test tubes contains 10 cm^3 of water at 25°C. The following masses of an unknown solid are placed in the test tubes: 4 g in the first, 8 g in the second, 12 g in the third, and 16 g in the fourth. After the tubes are shaken, it is observed that all of the solid has dissolved in the first two tubes, but that there is undissolved solid in the remaining two tubes.
a) What is the concentration of the solid in each of the first two tubes?
b) What can you say about the concentration of the solid in the second two tubes?

a) First tube: 4/10 × 100 = 40 g/100 cm^3
    Second tube: 8/10 × 100 = 80 g/100 cm^3
b) In the third and fourth tubes, the concentration is between 80 g/100 cm^3 and 120 g/100 cm^3.

26 a) Which of the substances shown in Fig. 4.3 could be the unknown solid of Prob. 25?
b) If the unknown is indeed the substance you named in answer to part (a), what will happen if the solution in each test tube is cooled to 10°C?

94 Solubility

a) Sodium nitrate

b) Nothing would be observed in the first tube. A small amount of solid (0.3 g) would precipitate out in the second tube. Some additional solid would appear in the third and fourth tubes. (There would be 4.3 g of undissolved solid in the third tube and 8.3 g of undissolved solid in the fourth.)

27 The solubility of a substance in water was found to be 5 g/100 cm^3 at 25°C, 10 g/100 cm^3 at 50°C, and 15 g/100 cm^3 at 75°C. What would you expect its solubility to be at 60°C? At 100°C? Explain.

To answer the questions, we draw a graph like that in Fig. IV and read the solubility at 60°C directly from the graph. It is 12 g/100 cm^3.

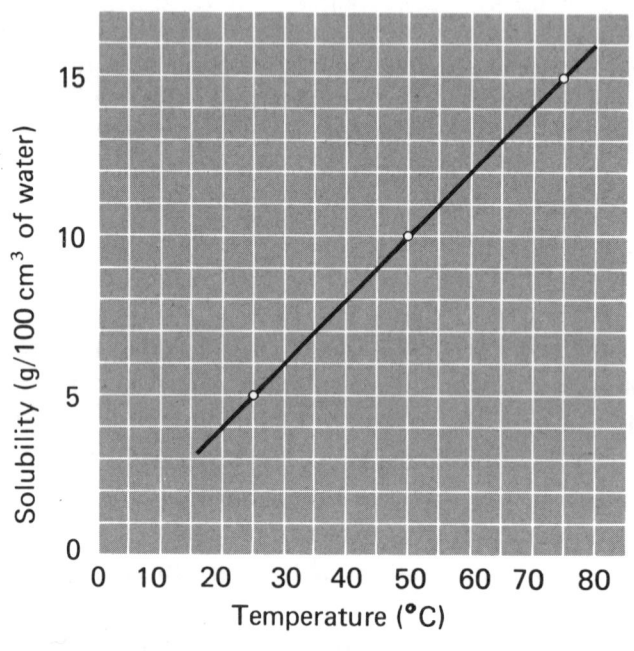

Figure IV

You can see that the solubility graph is a straight line in the temperature range for which the data are given. Therefore, we can extrapolate the graph to 100°C, assuming that it continues to be a straight line, to find the solubility at 100°C. This gives 20 g/100 cm^3. Such an extrapolation is no more than a reasonable guess, and we should not have as much confidence in our answer for 100°C as for 60°C. For example, if we place a straightedge along the solubility curve for sodium nitrate in Fig. 4.3, it appears to be a straight line from 0°C to 35°C, but extrapolating this curve as a straight line above 35°C would lead to increasingly greater errors in determining the solubility of sodium nitrate at higher and higher temperatures.

Solubility   95

28 In many localities, after a kettle has been used for some time for boiling water, a flaky solid appears on the inside-bottom and on the sides of the kettle where the water has been in contact. How do you account for the presence of this so-called "boiler scale"?

The water contained a small amount of dissolved solid; natural water is frequently saturated with minerals of small solubility that the water has dissolved out of the ground. The reduction in volume of the water as it boils away causes some solid to precipitate. As more water is added, the same thing occurs and the deposit of boiler scale builds up as the process is repeated.

29 Your experiment with moth flakes (Expt. 4.6) showed that this substance was insoluble in water but dissolved readily in methanol.
a) Predict the effect of adding water to a methanol solution of moth flakes. Try it.
b) Sugar proved to be almost insoluble in methanol but dissolved readily in water. Predict the effect of adding methanol to a solution of sugar in water. Try this too.

a) Students may predict that the moth flakes will remain dissolved because of the presence of methanol. Another possible prediction is that the moth flakes will come out of solution. In fact, the latter prediction is the correct one.
b) A laboratory test will show that the addition of methanol to the sugar-water solution does not result in a precipitate. This is a good example of the danger of generalizing on the basis of one experiment.
   Do not use burner fuel in place of methanol. It may give a cloudy solution when mixed with water.

30 In Expt. 4.8 you found that magnesium metal would dissolve in sulfuric acid.
a) Does this observation enable you to predict with certainty that all metals will dissolve in sulfuric acid?
b) Try dissolving other metals, such as copper, zinc, lead, and aluminum, in sulfuric acid.

a) You cannot predict from this observation whether all metals will dissolve in sulfuric acid. Each one must be tested in the laboratory.
b) If your students make these tests, they should wear safety goggles and be carefully supervised because of the dangers involved in using this acid. If dilute acid is used, the results are as follows: Zinc and aluminum will dissolve readily; copper and lead will not dissolve.

31 When magnesium carbonate is placed in sulfuric acid, a gas is produced whose properties you studied in Expt. 4.8. When you place

washing soda in hydrochloric acid, you also get a gas. What would you do with this gas to determine whether it is the same gas you got from dissolving magnesium carbonate in sulfuric acid?

We can test the gas to see (1) whether it is more or less dense than air, (2) whether or not it burns, and (3) whether or not it turns limewater milky. You might wish to have some students investigate this gas in the laboratory.

The gas obtained by dissolving magnesium carbonate in sulfuric acid is more dense than air, does not burn, and turns limewater milky. We get the same results from the gas obtained by putting washing soda in hydrochloric acid.

32 Three samples of gas are tested for characteristic properties. Sample *A* does not turn limewater milky, is less dense than air, and burns. Sample *B* turns limewater milky, is more dense than air, and burns. Sample *C* turns limewater milky, is less dense than air, and burns. What can you conclude about these samples of gas?

A table may be of help here.

Gas	Reaction with limewater	Density compared with air	Combustibility
A	No	Less	Burns
B	Yes	More	Burns
C	Yes	Less	Burns

Since no two of these gases have identical properties, they cannot be the same. Some students will try to identify the gases; for example, "*A* is hydrogen because we saw it burn, it is less dense than air, and it doesn't react with limewater." But *A* could equally well be a gas that the student has not yet seen—methane, for example. (In fact, all three could be gases not investigated by your students.)

Gas *B* might be a mixture of carbon dioxide and hydrogen. Gas *C* might also be a mixture of carbon dioxide and hydrogen, but with a greater proportion of hydrogen than gas *B*.

33 A fizzing tablet is dissolved in 10 cm^3 of water, and the gas is collected as in Expt. 3.9. The volume of gas collected is 450 cm^3. When 50 cm^3 of water is used, the volume of gas collected is 405 cm^3. The tube was all the way up in the bottle in both cases.
a) Why do you think the volume is less?
b) Would this make a difference in the density calculation?

a) The gas produced by the tablet must be somewhat soluble in water. When more water is used in the test tube, less gas is collected in the bottle. Therefore, more gas must be dissolved in the water in the test tube.

b) We measure only the mass and the volume of the gas that leaves the test tube. Thus, there will be no change in the measured density.

34 If you have a certain amount of a solid to dissolve in water, you usually can hasten the process by (a) stirring the water, (b) crushing the solid into smaller particles, and (c) heating the water. Why do you think each of these steps is effective in making the solid dissolve faster?

a) The water must come in contact with the substance to be dissolved. If the substance settles to the bottom of the container, the water can only contact the top layer. Stirring allows the water to contact more of the substance. In addition the water near any piece of the substance will begin to approach a saturated solution so stirring keeps the more dilute solution in contact with the substance.
b) Crushing the solid increases the area of the solid in contact with the water. You can see that this is so by thinking of a rectangular block of solid 1 cm × 1 cm × 2 cm; its total surface area is 2(1 cm × 1 cm) + 4(1 cm × 2 cm) = 10 cm². If you cut the block into two equal pieces across its long edge, you will have two cubes 1 cm on an edge; their total surface area is 6(1 cm × 1 cm) + 6(1 cm × 1 cm) = 12 cm². The finer the particles into which the solid is crushed, the greater the surface area in contact with the water.
c) The closer a solution is to being saturated with a given substance, the slower the substance will dissolve. The solubility of most solids increases with temperature. Heating the water makes the solution less saturated, so not only can the water dissolve more of the solid, but the dissolving will proceed faster.

The Separation of Substances

5

General Comments

As we have mentioned earlier, one of the criteria for selecting characteristic properties for discussion was their usefulness in separating substances. Now we employ these properties for actual separations in the laboratory, describe some extensions of these methods on an industrial scale, and arrive at an operational definition of a pure substance. Reading through this chapter, you may get the impression that we are leaving students with a rather vague definition of a pure substance. This is true. The boundary between a mixture and a pure substance is not so sharp as may be believed from reading some elementary textbooks. If your students realize at the end of this chapter that a pure substance is something that cannot be broken up by any of the methods discussed, they will have learned their lesson.

The schedule for this chapter is:

Sections 1-3 (one experiment, probs. 1-6, 20-23)	7 periods
Sections 4-7 (three experiments, probs. 7-14, 19, 24-28, 33	6 periods
Sections 8-10 (probs. 15-18, 29-32)	2 periods
Total	15 periods

EXPERIMENT: FRACTIONAL DISTILLATION 5.1

This method of separating substances is similar to the method used in the last part of the Distillation of Wood experiment in Chap. 1. There, students used distillation to separate a watery liquid from a dark, tarry residue that had been produced from wood. The purpose of this experiment is to look more closely at the distillation of a mixture of liquids.

A temperature-time graph made from readings taken during a trial distillation of 5 cm^3 of the liquid helps students decide on the number of fractions to collect. They then distill a larger sample and investigate the properties of the fractions. They should be sure to collect the various fractions in different collection test tubes over suitable temperature ranges chosen by examining the boiling-temperature graph.

The entire experiment will require at least five full class periods, but the time is well spent. Students should be allowed to work at their own speed after the trial distillation; however, they should also be encouraged to look ahead and to decide at what point to discontinue the experiment until the next period. It is most convenient to complete the distillation of the 25-cm³ portion of the solution within one period, and students should plan accordingly. Well-organized records of each day's work and frequent discussions with each group will help to keep things organized.

The Experiment

The solution used in this experiment is a mixture of isopropyl alcohol and water. For a class of 24 students (12 pairs), make up a mixture of 200 cm³ isopropyl alcohol and 300 cm³ water. Each pair of students will then have about 40 cm³, with a little left over.

Part A

In testing the original liquid (and later the three fractions), insist that your students measure density carefully. If they do not, the results will be inconclusive.

Suggest that students try the flammability test using dry paper, then paper moistened with water, then paper moistened with burner fuel, in order to be able, in Part F, to compare the results with those obtained in their tests of the original liquid (and various fractions).

To test sugar solubility, it is convenient to add about ¼ cm³ of sugar to about 1 cm³ of liquid. Students very often use too much sugar or do not wait long enough for it to dissolve. Thus, their results can be inconclusive.

Part B

When the first 5 cm³ is distilled, if there is any residue from Expt. 1.1 in the tubing, or if the rubber tubing is new, the alcohol will dissolve some material and the distillate will be yellow. Rinsing the tubing beforehand with alcohol will reduce this discoloration.

The solution will "bump," and you should have a supply of fresh boiling chips for your students. Since boiling chips cannot be used again, new chips must be added for the second distillation or if the distillation has been stopped. If you are out of boiling chips, broken porous porcelain is satisfactory.

In this experiment we are interested in the temperature of the condensing vapors as they leave the distilling test tube, not in the temperature of the liquid. Therefore, the thermometer bulb should be at the top of the test tube near the exit hole, not in the liquid as in Expt. 3.4. A thermometer cannot easily be inserted into and removed from the holes in standard No. 4 two-hole stoppers. In order to reduce thermometer breakage, and for safety, enlarge the hole with a file or drill. A few drops of glycerin on the thermometer stem makes it much easier to insert the thermometer.

During the distillations, the end tubing should be below the level of the water surrounding the collecting test tube. So that no part of the distillate can be forced back into the distilling tube, the level of liquid in the collecting test tube should not rise above the end of the tubing.

The best separation of fractions will occur if heating is carefully controlled and the temperature raised slowly. The distillation should be started with the test tube being heated through the wire screen by an alcohol burner. In order to keep the solution boiling vigorously (but not so strongly that it boils over into the receiving tube), the height of the test tube above the wire screen can be adjusted up or down to decrease or to increase the heat supplied. When distillation of the water begins to take place, the wire screen can be removed and the test tube heated directly with the burner.

A typical graph of temperature *vs.* time obtained from the distillation of a 5-cm^3 sample is shown in Fig. I. Give your students time to decide for themselves when to change collecting tubes to separate the various fractions, and then discuss with them the reasons for their decision before going on to the rest of the experiment. It is very important that a good post-lab class discussion be held at this point to insure that the students understand the process of distillation and the procedure that must be followed in the next step. From this point on, the class should be allowed to work at their own pace.

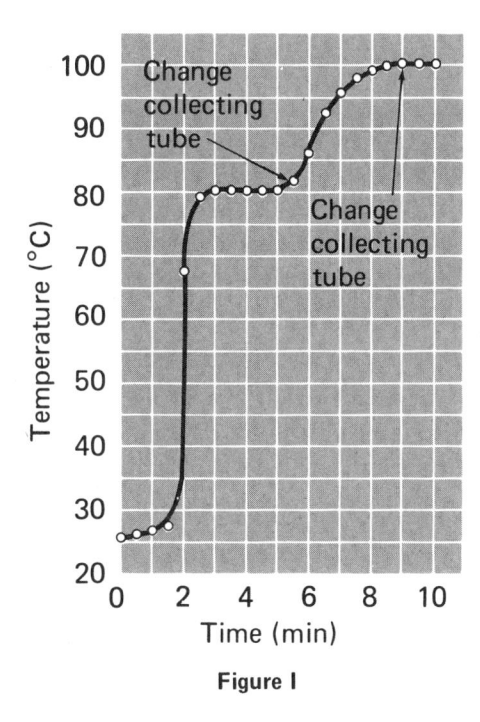

Figure I

Part C

When the experiment is performed carefully, the constant temperature of the plateau in the trial run will agree very closely with the boiling point of the fraction collected at the plateau temperature. However, this should be verified by plotting a boiling-point graph (this time with the thermometer in the liquid) for each fraction to see if a constant temperature plateau is obtained as the liquid is boiled away nearly to dryness. The second distillation (25 cm³ of solution) gives enough of each of three fractions for performing the required tests after the boiling-point graph is obtained.

Little distillate condenses until the temperature reaches about 78° to 79°C, and thus Fraction No. 1 is collected from about 78° to 82°C.

Fraction No. 1, which will contain about 9 cm³, has a very constant boiling point but is actually an azeotropic mixture consisting of 90 percent alcohol and 10 percent water. This unique mixture has a constant boiling point of 80.5°C and therefore cannot be separated by distillation. For this reason the azeotropic mixture has the appearance of a single substance in this experiment. The important point for the discussion of the experiment is that a separation has taken place, as the boiling points and other characteristic properties will show. The purity of the substance is not our concern here.

Fraction No. 2, about 4 cm³, will consist of a middle sample collected between about 82°C and 98°C.

About 12 cm³ of water will be collected as Fraction No. 3.

Parts D, E, and F

Sample Data

Property	Solution	Fraction 1	Fraction 2	Fraction 3
Flammability	Yes	Yes	Yes	No
Odor	Medium	Strong	Medium	None
Density	0.91 g/cm³	0.84 g/cm³	0.93 g/cm³	1.03 g/cm³
Boiling point	Not constant	80°C	Not constant	100°C
Dissolves sugar?	Yes	No	Slightly	Yes

The boiling-point graphs of the three fractions are shown in Fig. II. It is a good idea, in the post-lab, to make histograms of densities and boiling points from class data for the three fractions and the original liquid.

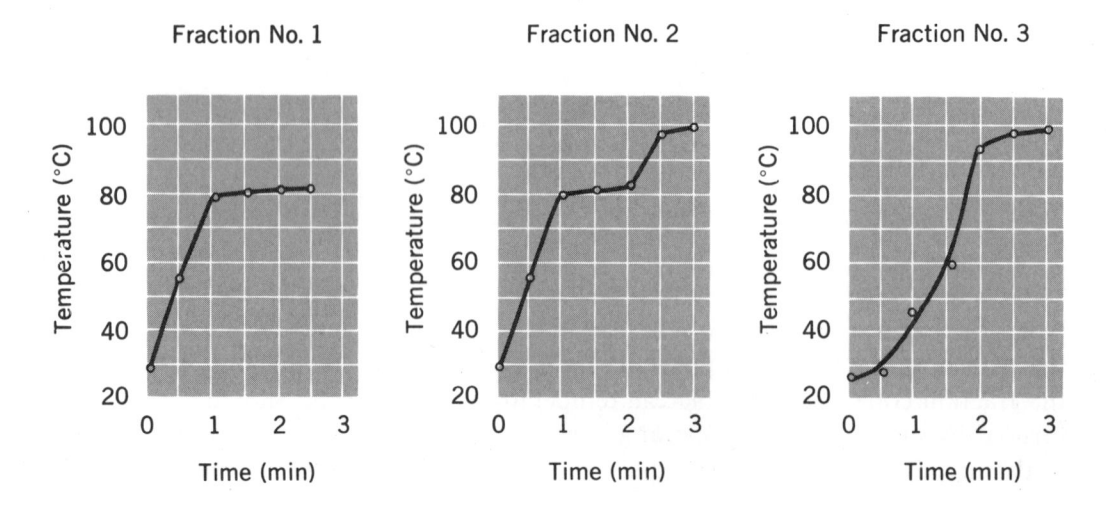

Figure II

The Separation of Substances   103

Answers to Questions

Part A. It is not possible to tell, just by looking, that the liquid is a mixture. The odor is somewhat like that of cleaning fluid. The liquid does burn if it is absorbed by a piece of paper and then lighted with a match. Its density is about 0.91 g/cm^3. Sugar dissolves fairly readily in the liquid.

Part B. The graph, which has been plotted from the boiling-temperature data, shows clearly two level sections or plateaus. This suggests that at least two different substances are present in the liquid and that they can perhaps be separated by collecting three fractions. The temperatures at which collecting tubes should be shifted are about 82°C and 100°C. (See Fig. I.)

Part C. The odor and the flammability of each fraction are given under Sample Data.

Parts D and E. The density, boiling point, and solubility of sugar in each fraction are listed under Sample Data.

Part F. See Sample Data. The composition of the fractions is as follows: Fraction No. 1 appears to be a pure substance with a boiling point of about 80°C; Fraction No. 2 is a mixture of substances similar to the original solution; Fraction No. 3 is a pure substance with a boiling point of about 100°C, probably water. From Table 3.2 it is possible that the lower-boiling liquid (boiling point 80°C) is isopropanol or *t*-butanol. The higher-boiling liquid (boiling point 100°C) could be either *s*-butanol or water.

Another test which would be helpful in positively identifying these substances is the measurement of freezing point. Table 3.2 shows that *t*-butanol would freeze at room temperature (melting point 26°C), while water freezes at 0°C.

If each of the fractions is redistilled into three equal volumes, there is no change in Fractions No. 1 and No. 3. Fraction No. 2 should separate into new fractions similar to those obtained from the original solution.

Apparatus and Materials

Pegboard
Balance
Beaker (250 cm^3)
Large test tube
6 Small test tubes
Rubber or plastic tubing (25 cm)
Burner fuel
3 No. 2 solid stoppers
Two-hole stopper (No. 4, with 1 hole large enough to hold a thermometer)
Isopropyl alcohol-water mixture (40 cm^3 per pair of students)
Boiling chips
Sugar (5 g)
Thermometer
Large clamp
2 Small clamps
Test-tube rack
Burner
Burner stand
Short bend of glass tubing
Matches
Marking pencil for glass
Graduated cylinder (10 cm^3)
Paper towels
Plastic container for density measurement
Graph paper
Water
Glycerin
Safety glasses

104   The Separation of Substances

5.2 PETROLEUM

This section can be treated as a reading assignment. The study of petroleum is discussed as an example of the practical application of fractional distillation in separating substances. It is not expected that students should specifically learn the fractions of petroleum or the various geological changes and formations.

5.3 THE DIRECT SEPARATION OF SOLIDS FROM LIQUIDS

If the ideas in this section are not clear to your students, you should do a simple demonstration in which a mixture of sawdust, sand, and water is separated.

The technique of filtering and then boiling the filtrate to dryness is used in the next experiment. Use the discussion of this section as an introduction to the next.

5.4 EXPERIMENT: SEPARATION OF A MIXTURE OF SOLIDS

The easiest way to separate a mixture of two solids, other than by a mechanical method, is to find a solvent that will dissolve one and not the other. This method is employed in this experiment to separate a mixture of salt and sulfur. Both the understanding and the manual skills acquired in this experiment serve as preparation for the next one, in which similar methods are used in a more sophisticated way.

The Experiment

Enough of the mixture of sulfur and sodium chloride for 24 students (12 groups) can be made by mixing 5 g of sulfur with 15 g of sodium chloride. Give each group about 1.5 g of the mixture. If the entire mixture is ground up with a mortar and pestle and mixed well, it is very difficult to detect visually the presence of more than one substance.

In the pre-lab, show your students how to fold the filter paper to form a cone. If the paper is wet when placed in the funnel, it will fit snugly against the walls of the funnel.

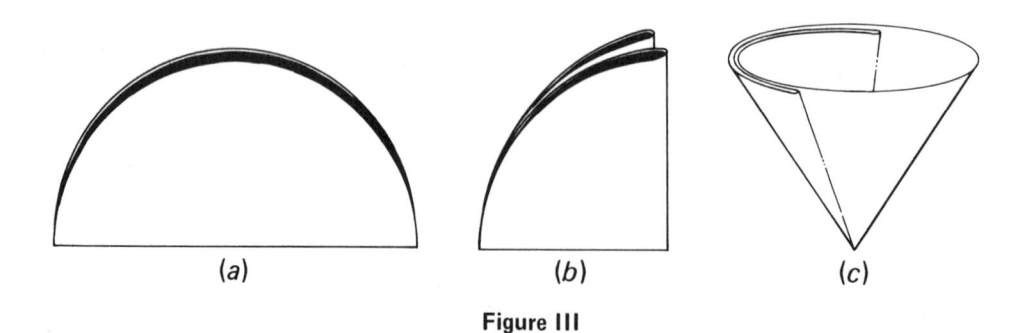

(a)   (b)   (c)

Figure III

The method of folding the paper is illustrated in Fig. III. It is first folded in half as shown in (*a*), then in half again as shown in (*b*). Finally, it is opened into a cone, as shown in (*c*), dampened, and gently fitted into the funnel. Warn your students not to push it into the funnel with a finger (or pencil) against the bottom (apex of the cone).

Answers to Questions

The mixture is quite impossible to separate with tweezers, but the substances can be clearly separated by the method used here.

Apparatus and Materials

Sulfur and sodium chloride mixture (1.5 g)
2 Test tubes and stopper
Test-tube rack
Filter funnel
Filter paper
Evaporating dish
Burner and burner stand
Safety glasses

THE SEPARATION OF A MIXTURE OF SOLUBLE SOLIDS 5.5

Fractional crystallization is one of the most widely used methods of separating and purifying substances. This method will be used in a simplified form in the next experiment to separate a mixture of sodium chloride and potassium nitrate. Because experience has shown that many students have difficulty in following the reasoning involved in fractional crystallization, we devote this entire section to the rationale for the procedure to be used in Sec. 5.6.

If you have a strong class we suggest that you take the time to go over the material with your students in detail. If you have a weak class you may prefer to skip Secs. 5.5 and 5.6 entirely.

Problem 12 should be assigned and discussed as a preliminary; and Problems 9-11 can be assigned to your more able students immediately after completing Expt. 5.6 to reinforce what has been studied in the section. Problems 13 and 14 suggest another way to perform the separation: by boiling away solution and then filtering while hot. In fact, the procedure described in these problems is closely related to the answer to Problem 11.

EXPERIMENT: FRACTIONAL CRYSTALLIZATION 5.6

The purpose of the experiment is stated in the text: to test the predictions made in the previous section, by using the same quantities of solids and the same amount of water as determined in that section.

The Experiment

For a class of 24 students, thoroughly mix 70 g of potassium nitrate and 65 grams of sodium chloride. Because common salt has an additive that forms a cloudy solution, use reagent grade sodium chloride. Give 10 g of the mixture to each pair of students; they should verify with their balances that they have received 10 g of the mixture. They need not worry if their sample is a few tenths of a gram too heavy or too light.

All apparatus should be assembled in advance of starting the experiment so that the funnel and filter paper can be preheated.

A convenient way to heat the solution is to put the test tube containing it into a beaker of water heated to a temperature above 70°C. When the temperature of the solution is well above 70°C, the hot water in the beaker can be poured through the filter paper and funnel to heat them. The hot water can then be discarded. Now the solution can be poured into the filter paper in the funnel and the filtrate collected.

Part of the experiment is to prepare crystals of sodium chloride. It would be well to do the same for some potassium nitrate fresh from the storage bottle. This makes it possible to observe that potassium nitrate is, in fact, recovered in the end.

Answers to Questions

You must be sure that the temperature is 70° (or perhaps higher); 3.6 cm^3 of water was picked because at this temperature the water would dissolve all of the potassium nitrate, and at a lower temperature it will not do so.

You expect undissolved sodium chloride on the filter paper. You expect potassium nitrate to precipitate from the filtrate and practically no sodium chloride.

The crystals that precipitated from the filtrate as it cooled are large, regular, and needle-shaped.

The crystals of sodium chloride are in the shape of small cubes.

Another way that you could show that the two solids that separated are different is by comparing their solubilities. You could put equal masses of the two solids in separate test tubes, then add hot water, a few drops at a time, until all of the solid had dissolved. You would observe that much more *hot* water is required to dissolve the sodium chloride than the potassium nitrate.

Apparatus and Materials

Balance
Pegboard and clamps
Beaker (250 cm^3)
2 Test tubes (20 × 150 mm)
Test-tube rack
Burner
Burner stand
Graduated cylinder (10 cm^3)
Stirring rod
Watch glass

Thermometer
Funnel
Filter paper
Scoopula
Potassium nitrate (5 g)
Sodium chloride (5 g)
Water
Paper towel
Safety glasses

5.7 EXPERIMENT: PAPER CHROMATOGRAPHY

This experiment illustrates another method of separation that has been used in biochemical research. Paper chromatographic separation is easy to do, but a lot more difficult to explain than, say, fractional distillation. We suggest that you be satisfied with using it as a tool.

This experiment can be performed at home, using a large milk bottle or other tall transparent container in place of a graduated cylinder. If filter paper is not available, a blotter, paper towel, or paper napkin can be used to perform the experiment, though these materials do not work as well as filter paper.

The ink mark is most easily drawn by a single stroke of a small water-color paintbrush. Letting the ink dry on the paper will not affect the quality of the chromatograph. In fact, drying often improves the chromatograph.

The Experiment

Each of the colors making up the ink represents a different substance, and each substance travels up the paper at a different rate. Therefore, the farther the ink has to move, the greater the separation that will be observed in the individual colors. Best results are obtained when a large amount of ink is deposited over a small area of the strip of filter paper. Although a single streak is sufficient, it is better to allow this to dry and cover it with a second or third streak.

The water carries the ink slowly up the filter paper, and it is important to have the top covered with aluminum foil to prevent evaporation of the water. A glass plate may also be used as a cover.

The time required for the fastest-moving fraction of the ink to come to within 2 cm of the top is about 30 minutes.

The filter paper should not come into contact with any moisture on the walls of the container. Using a larger container, such as a milk bottle, will prevent this from happening. A taller container gives a longer chromatograph with better separation of the colors.

After the paper strip has dried overnight, it is a simple matter for students to cut out the colored sections with scissors. Each section should be cut into smaller pieces, then put into a test tube and barely covered with a little water. About 0.5 cm^3 to 1.0 cm^3 of water is enough; the less used, the more concentrated the solution will be. The color can be extracted more rapidly by mashing the pieces in the water with a glass stirring rod or a narrow wooden splint.

Because of dilution, the color will most likely be a gray-blue instead of black when the liquids from all the test tubes are combined. However, the combined color can be compared with that of the original ink by diluting a small amount of the latter with water in a test tube. A solution of ink as black as the original could be obtained only by combining all the different-colored fractions and concentrating them by distilling off excess water.

You may find it best to collect all the cut up colored pieces from the whole class and reconstitute the black ink as a class demonstration. There is more information on chromatography in *The Amateur Scientist*, by C. L. Stong (Simon and Schuster, 1960).

Students who are interested in individual work should be encouraged to make chromatograms using various inks, dyes, etc., and solvents such as alcohol, acetone, etc., in place of the water.

Answers to Questions

The number of substances that can be identified by chromatography will depend on the degree of separation and the kind of ink used. Carter's ink gives blue, purple, and orange bands. Sheaffer's ink gives blue, yellow, and purple bands. Where the blue and yellow overlap, they will appear as green, and often only a very little yellow can be detected.

The colors can be extracted from the filter paper again by dissolving in water, and these colored solutions recombined to give a blue-gray solution, which compares closely with a diluted sample of the original ink.

Apparatus and Materials

Small test tube
Test-tube rack
Graduated cylinder (50 cm^3)
Filter paper strip (20 cm long)
Small paintbrush
Aluminum foil (6 cm sq.)
Scissors
Stirring rod

Ink, Carter's washable black, Sheaffer Skrip washable black, or Parker's Quink permanent black (5 cm^3 for class)
Water
Paper towel

5.8 MIXTURES OF GASES: NITROGEN AND OXYGEN

Air was considered to be a single substance until the end of the eighteenth century, when its components were recognized. Note that the properties of air, a mixture, are a compromise between the properties of its components. The density of air is between the density of nitrogen and the density of oxygen. A splint glowing in air goes out in nitrogen and bursts into flame in oxygen.

5.9 LOW TEMPERATURES

The purpose of this section is threefold: first, to put some meaning behind the low temperatures quoted for the boiling points of oxygen and nitrogen; second, to illustrate that the experimental tools used in this course cover only a minute fraction of the wide range available to science; and finally, to give students some practice in grasping the essentials of an experimental situation merely by reading a description. Here we have chosen a situation which is an extension of what was done in the laboratory.

A simple gas thermometer, as in Fig. 5.8, can be used only for a limited range—that in which the liquid drop remains a liquid. For very-low-temperature work the apparatus can be adapted as follows: A bulb containing a gas is connected to the scale part of the thermometer by a capillary tube. The bulb is placed in the low-temperature region, while the liquid drop in the scale is maintained at a temperature above its freezing point.

MIXTURES AND PURE SUBSTANCES 5.10

This section summarizes the whole chapter and much of the course up to this point. It also leads naturally into Chap. 6. Treat it very thoroughly. If necessary, read it in class, or ask your students to write a summary of the main ideas.

Note that the operational definition of a pure substance is not a tight one. There is no need to apologize for this. This is the way science works.

110   The Separation of Substances

CHAPTER 5—ANSWERS TO PROBLEMS

Sec.	Easy	Medium	Hard	Class Discussion	Home or Lab
1	1†, 2, 23	3		2, 3	
2	22	4, 20, 21			
3	5†, 6†				
4	7†, 8, 19	24, 25			19
5-6		9, 10, 11, 12, 13, 26	14	9, 10, 11, 12, 14	
7		28		28	27
8	29	15			
9		16†	30		
10	17	18, 31, 32, 33			

1† In what characteristic property must two liquids differ before we can consider separating a mixture of them by fractional distillation?

They must have different boiling points.

2 The temperature-time graph shown in Fig. A was made during the fractional distillation of a mixture of two liquids A and B, and fractions were collected during the time intervals I, II, III, and IV. Liquid A has a higher boiling point than liquid B. What liquid or liquids were collected during each of the time intervals?

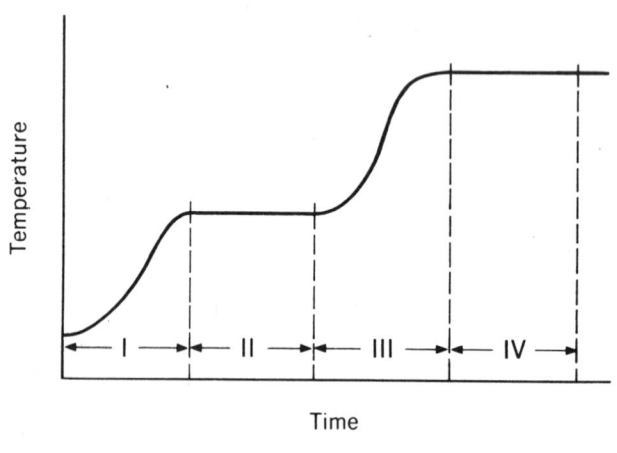

Figure A

Interval I: Very little liquid is collected during this interval, and it is almost pure liquid B.
Interval II: Almost pure liquid B is collected during this interval.
Interval III: A mixture of liquids A and B is collected.
Interval IV: Almost pure liquid A is collected.

3 Carlos boiled a liquid and recorded the temperature at 1-min intervals until the liquid had nearly boiled away. How do you explain the shape of the curve he got? (See Fig. B.)

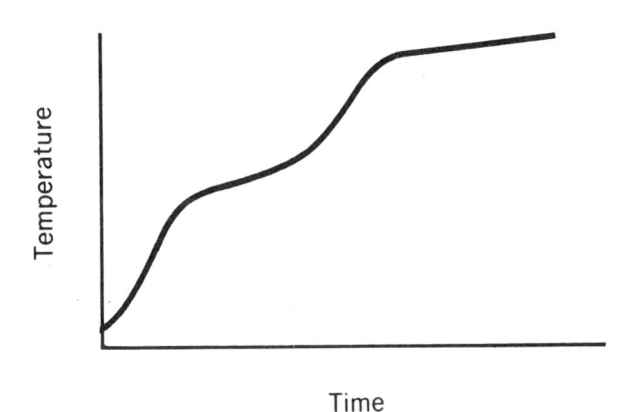

Figure B

The purpose of this question is to make students aware that a mixture of liquids, when distilled, does not always give flat plateaus such as the ones they obtained in Expt. 5.1. There are several possible reasons why flat plateaus may not be obtained. The liquid may be a mixture of many substances, some of which have boiling points very close together. If the liquid is heated too rapidly, a series of short plateaus will appear as a rising curve. The liquid also may be composed of only two substances that do not separate easily. A mixture of methyl alcohol and water is an example.

4 A sample of crude oil is boiled for several minutes. What change takes place in its density? (See Table 5.1.)

The density increases.

5† How could drinking water be obtained from seawater?

By distilling the seawater.

6† Suppose you have a barrel filled with a mixture of sand, small pebbles, and stones. How would you separate the components of this mixture?

By using two sieves. One sieve allows pebbles and sand to go through but keeps back the larger stones. The other sieve allows only the sand to go through, while keeping back the pebbles.

7† In what characteristic property must two solids differ if they are to be separated merely by dissolving at room temperature and filtering?

They must differ greatly in solubility.

8   Much salt is obtained from salt mines, in which great masses of salt occur mixed with insoluble earthy impurities. What steps could be taken to purify the salt?

The salt can be dissolved in water. The mixture can be filtered to separate out the earthy impurities, and then the salt can be recovered by evaporating the water.

9   a) How much sodium chloride does not dissolve in the 3.6 cm^3 of water at 70°C?
    b) How much sodium chloride remains in solution after cooling to room temperature?
    c) How much solid sodium chloride is recovered at the end of the experiment?

a) From Fig. 5.4, 38 g of sodium chloride dissolve in water at 70°C, and so $38 \text{ g} \times \frac{3.6 \text{ cm}^3}{100 \text{ cm}^3} = 1.4$ g of sodium chloride dissolve. Hence $5.0 - 1.4 = \underline{3.6 \text{ g}}$ do not dissolve.
b) Take room temperature as about 20°C. From Fig. 5.4, the solubility of sodium chloride at room temperature is about 36 g/100 cm^3 of water. In 3.6 cm^3 of water, $36 \times \frac{3.6}{100} = 1.3$ g remain dissolved.
c) $5.0 \text{ g} - 1.3 \text{ g} = 3.7$ g of sodium chloride is recovered on the filter paper.

10 a) How much potassium nitrate remains in solution at room temperature at the end of the separation?
    b) How much potassium nitrate can be recovered in solid form?

a) Again take room temperature as about 20°C. The solubility of potassium nitrate at room temperature is about 32 g/100 cm^3 of water. In 3.6 cm^3 of water, $32 \text{ g} \times \frac{3.6 \text{ cm}^3}{100 \text{ cm}^3} = 1.2$ g of potassium nitrate remain dissolved.
b) Thus $5.0 \text{ g} - 1.2 \text{ g} = 3.8$ g of potassium nitrate precipitated on cooling the filtrate and can be recovered as a solid.

11 Suppose you wish to separate more of the two solids remaining in solution at the end of the experiment. How would you proceed to do this?

At the end of the experiment, you have 3.6 cm^3 of water containing 1.3 g of sodium chloride and 1.2 g of potassium nitrate. You can heat the solution until it boils, and then boil the solution until its volume has been reduced to, say, 1.0 cm^3. In this volume of water at 100°C (the boiling temperature is actually higher than this), 0.4 g of sodium chloride and 2.4 g of potassium nitrate could dissolve. Therefore all the potassium nitrate remains dissolved, but 1.3 g − 0.4 g = 0.9 g of sodium chloride would precipitate. Once again you filter the solution and let it cool to room temperature. 1.0 cm^3 of water at room temperature will dissolve 0.4 g of potassium nitrate and 0.4 g of sodium chloride. Almost all of the sodium chloride would remain in solution, but 0.9 g of potassium nitrate would precipitate on cooling.

12 a) If a solution containing 40 g of potassium nitrate in 100 cm^3 of water at 100°C is cooled to 25°C, how much potassium nitrate will precipitate out of solution? (See Fig. 5.4.)
b) Suppose that the 40 g of potassium nitrate were dissolved in only 50 cm^3 of water at 100°C. How much potassium nitrate will precipitate out if the solution is cooled to 25°C?

a) Figure 5.4 gives the solubility of potassium nitrate as 38 g per 100 cm^3 of water at 25°C. Thus, 40 g − 38 g = 2 g of potassium nitrate will precipitate.
b) Since 38 g of potassium nitrate will remain dissolved in 100 cm^3 of water at 25°C, only half as much, 19 g, will remain dissolved in 50 cm^3 of water at this temperature. Thus, 21 g will precipitate when the solution is cooled to 25°C.

13 Suppose you dissolve 30 g of sodium chloride in 100 cm^3 of water at 100°C and boil away 50 cm^3 of the water.
a) How many grams of sodium chloride will remain in solution?
b) How many grams will precipitate out of solution?

a) Figure 5.4 shows the solubility of sodium chloride to be 40 g per 100 cm^3 of water at 100°C, or 20 g per 50 cm^3. Thus, 20 g of sodium chloride will remain in solution in 50 cm^3 of water at 100°C.
b) 30 g − 20 g = 10 g of sodium chloride will precipitate.

14 Suppose you dissolve 40 g of potassium nitrate in 100 cm^3 of water at 100°C.
a) If half the solution is poured out, how many grams of potassium nitrate will the remaining solution contain?
b) Now, instead of pouring out part of the solution, you boil away 50 cm^3 of water. How many grams of potassium nitrate will remain in solution at 100°C?
c) If the solution remaining in (b) were cooled to 25°C, how much potassium nitrate would precipitate out of solution?

a) The remaining half of the solution will contain 40 g/2 = 20 g of potassium nitrate.
b) Figure 4.3 shows the solubility of potassium nitrate to be 241 g per 100 cm^3 of water, or 241 g/2 = 121 g per 50 cm^3 at 100°C. Only water is boiled away, and thus all of the original 40 g of potassium nitrate will remain in solution in 50 cm^3 of water at 100°C.
c) Figure 4.3 shows that the solubility of potassium nitrate is 38 g per 100 cm^3 of water at 25°C, or 19 g per 50 cm^3. Thus 40 g − 19 g = 21 g of potassium nitrate will precipitate when the solution in (b) is cooled to 25°C.

15 **How could you separate ammonia gas from air?**

The two gases can be separated by bubbling the mixture through water. The ammonia will dissolve in the water and can be recovered by heating the water. Being only slightly soluble, the air will bubble up through the water and can be collected by water displacement.

Another method is to cool the gases. Ammonia will liquefy at about −34°C and can be drained off into a separate container.

16† **Liquid nitrogen will boil in a teakettle resting on a cake of ice. How do you account for this?**

The temperature of the ice (0°C) is above the boiling point of liquid nitrogen (−196°C) as given in Table 5.2.

17 **Suppose you mixed together all the fractions you obtained from the fractional distillation of the liquid in Expt. 5.1. What do you think would be the properties of this liquid?**

When the fractions obtained from the fractional distillation of a liquid are recombined, the mixture of fractions will have the same properties as the original liquid. Each fraction will contribute partially to the overall behavior of the mixture according to the amount present.

18 **The substances you obtained by distilling wood (Expt. 1.1) when mixed together will not give anything like wood—even ground-up, finely powdered wood. What does this tell you about the substances in wood as compared with the substances you obtained in the distillation?**

The substances obtained from the fractional distillation of a liquid, when mixed together, give back the original mixture, indicating that there were no changes in the substances in the original liquid. The fact that the products of the distillation of wood cannot be mixed together to give wood means that at least some of these products are new substances with properties quite different from the substances making up wood.

19 **In earlier times people would search for and find sandy stream beds in which small particles of gold were mixed with the sand. They separated this gold from the sand by "panning." Find out how this was done. What characteristic property of the substances made panning possible?**

One source of information is an article in the American Heritage Junior Library entitled "The California Gold Rush."

The sand and the gold are placed in a flat pan, like a pie pan, with some water. The mixture is swished around while the pan is tilted a little so that some of the water spills out. The moving water carries away the lighter sand grains, whereas the much denser gold settles at the bottom. Water is added as necessary. The process succeeds because of the great difference in density between the sand and the gold.

20 Figure 5.2 shows four oil wells drilled into oil-bearing porous rock. Can you suggest some method, without drilling deeper, for getting more oil from well *D* after the oil level drops below the end of the well?

If water is forced down into the well, the level of water trapped in the porous rock will be raised, and so will the layer of oil on top of it. In this way, the oil reaches the end of the well and can be pumped to the surface. Since the oil layer continues to rise as more water is forced in, almost all the oil can be taken from such a well before water begins to be pumped up.

21 a) In Table 5.1 (page 71), what fractions would be liquid at room temperature (20°C)? Which would be solids? Which would be gases?
b) You can see from the table that pentane is not an ingredient in any of the common products listed. How can you account for this?

a) Methane, ethane, propane, and butane are gases, and the remaining substances (pentane through hexadecane) are liquids at 20°C.
b) Since pentane has a boiling point of 36°C, it is not a gas at room temperature and cannot be included in the "fuel gas" fraction of petroleum. On the other hand, if pentane were included in the gasoline fraction, it would readily evaporate or boil out of the mixture during the hot summer months and probably cause a vapor lock in the fuel line if used in a gasoline engine. Pentane actually belongs in a category between the "fuel gas" and "gasoline." This category includes the highly volatile fractions of gasoline. It is used primarily for solvents and cleaning fluids when rapid evaporation is an advantage.

22 Using the data in Table 5.1 (page 71), draw and label a possible distillation curve for a mixture of hexane, nonane, and tetradecane.

The boiling points from Table 5.1 are: hexane, 69°C; nonane, 151°C; tetradecane, 254°C. The relative lengths of the plateaus would depend on the relative amounts of the three substances in the mixture.

23 When ethanol is to be used for commercial or industrial purposes other than as a beverage, it is customarily "denatured"; that is, a small quantity of another substance is added to it so that it is unfit

116   The Separation of Substances

for use as a beverage. What, in general, do you think some of the properties of the added substance might be?

The substance added will either be quite disagreeable to drink, cause severe gastric disturbances, or may be poisonous. It should mix freely with the ethanol, have a boiling point somewhere near that of ethanol so that it cannot be easily separated by distilling, and should not interfere with the intended industrial use of the ethanol. Methanol is frequently used to denature ethanol; there is a wide variety of other substances that are also used. Some of these are acetone, brucine, and tert-butyl alcohol; aviation gasoline is also used.

24 **How would you separate a mixture of powdered sugar and powdered citric acid?**

The substances can be separated by adding methanol to the mixture. All the citric acid will dissolve, while the relatively insoluble sugar will remain and can be filtered out. In order to obtain the citric acid again as a solid, the alcohol solution must be evaporated to dryness.

25 **The mineral called "Gay-Lussite" appears to be a pure substance, but it is actually a mixture composed of calcium carbonate (limestone) and sodium carbonate (soda ash) and water. Describe how you would go about separating these three substances from the rock. Some properties of calcium carbonate and sodium carbonate are listed in the following table.**

Property	Calcium carbonate	Sodium carbonate
Melting point	Decomposes at 825°C	851°C
Solubility in alcohol	Insoluble	Insoluble
Solubility in hydrochloric acid	Soluble	Soluble
Solubility in water	Insoluble	7 g/100 cm^3 at 0°C; 45 g/100 cm^3 at 100°C

The substance most easily separated from this mixture is the water. This is best done by breaking up the rock sample into small pieces and grinding them to a powder. The powder can then be placed in a test tube and heated to drive off the water, which can be condensed and collected in another test tube.

We see from the table that sodium carbonate is soluble in water whereas calcium carbonate is not. The dry solid could be boiled with a large amount of water for some time to allow the sodium carbonate to dissolve. The calcium carbonate, which remains solid, can be filtered off and collected on a filter paper. The dissolved sodium carbonate can be recovered by evaporating the clear filtrate to dryness. The dry residue contains no calcium carbonate.

The Separation of Substances   117

26 If you have 100 cm³ of water at 100°C, saturated in both potassium nitrate and sodium chloride, what happens if the temperature is lowered to 10°C? (Refer to Fig. 5.4, and assume that the solubility curves of these substances are the same as in the figure, even when they are dissolved together.)

From Fig. 5.4:

Solubility in Water (g/100 cm³)

Temperature (°C)	Sodium Chloride	Potassium Nitrate
100	40	241
10	36	22
	4	219

Four grams of sodium chloride and 219 g of potassium nitrate will precipitate.

27 You can use paper chromatography to separate the components in many common substances. Here are some you can work with at home: tomato paste, different colors and brands of ink, the coloring in leaves and vegetables (grind the leaves first in alcohol), and flower petals.

The strong coloring matter or pigments in plants and vegetables can usually be extracted with alcohol. However, the technique of many paper chromatography separations is quite complex. Interested students should refer to Chapter 5 of *The Amateur Scientist* (Simon and Schuster, 1960).

28 Chlorophyll can be extracted from leaves by grinding them with alcohol to give a dark-green solution. By careful application of paper chromatography, bands of yellow and red color, as well as green bands, can be detected. What other reason do you have to suspect the presence of substances producing these colors in leaves? Why don't you ordinarily see them?

Leaves turn red and yellow in the fall, when the amount of chlorophyll is diminished. At other times, the red and yellow substances present in the leaves are masked by the chlorophyll.

29 As liquid air boils away, the remaining liquid becomes richer in one of the two gases—nitrogen or oxygen. Which one is it? How do you know?

Nitrogen, having a somewhat lower boiling point than oxygen, boils off the mixture more readily, thus leaving the remaining mixture increasingly richer in oxygen.

118   The Separation of Substances

30 a) How would you calibrate the simple gas thermometer shown in Fig. 5.8 to read in Celsius degrees?
b) Which end of the liquid drop would you take as a reference point?

a) Immerse the thermometer in an ice-water mixture and mark the position of the liquid drop "0°C." The thermometer must be immersed so that all the trapped gas is below the surface of the liquid. Repeat this procedure in boiling water, and mark this position "100°C." Divide the distance between these two marks into 100 equal divisions.
b) Either end of the liquid drop can be used as a reference point, but the same end must be used for both calibration marks and for all measurements.

31 What would you do to separate (a) alcohol from water, (b) sodium chloride from sodium nitrate, (c) nitrogen from oxygen?

a) On the basis of the difference in boiling points, alcohol can be separated from water by distillation.
b) Sodium chloride and sodium nitrate can be separated by fractional crystallization, which makes use of their different solubilities in water.
c) The difference in boiling points is widely used to separate oxygen from nitrogen. Oxygen could also be separated from nitrogen by burning something in the mixture or by using steel wool, which would remove the oxygen in rusting, with no effect on the nitrogen.

32 A sample of a liquid was boiled for 12 min. During that time the boiling point remained constant and the volume was reduced to half. Is the liquid a pure substance?

The liquid may be either a pure substance or a mixture of two or more pure substances with different boiling points but with the liquid of lowest boiling point making up more than half the total volume.

33 Suppose you had a mixture of sand and salt in a small box.
a) How could you separate these substances?
b) How would you determine the ratio of the mass of sand to the mass of salt?
c) If you were mixing sand and salt together, what mass ratios would it be possible for you to make?

a) Place the mixture in water. The salt will dissolve and can then be separated from the sand by filtering.
b) After the sand has been dried and the water evaporated from the salt, the mass of the solids can be determined by use of a balance,

and the ratio can be computed from the masses. Alternatively, one could subtract the mass of sand from the original mass of sand plus salt to get the mass of salt. This would eliminate the need for evaporation.

c)   You can make any ratio you want, from a huge excess of salt to the other extreme of a huge excess of sand.

Compounds

6

Overview of the Chapter

By definition, pure substances do not break up into different components by the application of the same separation methods that can be used to separate mixtures. The aim of this chapter is to show that in general pure substances can, nevertheless, be broken up by other means, such as intense heating and the application of an electric current. Conversely, such pure substances (compounds) can also be synthesized from other pure substances, but only in definite proportions of the components.

Our first step is to decompose two pure substances by using heat (Expt. 6.1) and electricity (Expt. 6.2). In each case new pure substances are produced that are quite different from the starting substance. We then reverse our method of attack and synthesize compounds. The examples we use are specially chosen to illustrate one of the basic differences between compounds and mixtures—that is, that we can synthesize compounds only with definite proportions of the components.

The choice of the two examples—the synthesis of water in Sec. 6.3 and the synthesis of zinc chloride in Expt. 6.4—is far from arbitrary. Many reactions that one can think of and write out are actually very complex to perform. The reactions often proceed very slowly and reach completion after a long period of time. Many times, one or more additional products are formed during the reaction. Thus, the genius of Proust in formulating the law of constant composition with conflicting evidence should be properly appreciated.

Early difficulties in the formation of this law sprang in part from the difficulty of determining when a reaction was complete. As an illustration of this circumstance, the reaction between copper and oxygen is studied. The investigation as to what has happened leads simultaneously to an understanding of complete and incomplete reactions and serves as an introduction to the next chapter.

The suggested schedule for this chapter is:

Sections 1-5 (three expts., probs. 1-20)	10 periods
Sections 6-8 (two expts., probs. 21-24)	4 periods
Achievement Test No. 3	2 periods
Total	16 periods

6.1 EXPERIMENT: DECOMPOSITION OF SODIUM CHLORATE

After the students have examined the melting and solubility of the initial and final solids and the flammability of the gas produced in this experiment, they determine that they are unable to produce the starting substance by remixing the products. They then conclude that they have decomposed a compound and not just separated substances from a mixture.

The Experiment

CAUTION: Sodium chlorate, though not dangerous if used as directed, is a strong oxidant and may cause a fire or an explosion when heated in the presence of any combustible material. Be sure you supply a glass-wool and *not* a cotton plug for insertion in the top of the test tube to prevent contact of hot sodium chlorate with either the rubber stopper or the rubber tubing. Do *not* use a catalyst with the sodium chlorate.

As the sodium chlorate is heated with two alcohol burners (or one Bunsen burner), the solid melts and the liquid begins to bubble as if it were boiling. Actually, as it changes to sodium chloride, the molten sodium chlorate is giving off oxygen, which will be collected in test tubes over water and identified. With continued heating, the liquid bubbles and foams vigorously, rising ½ to ⅔ of the way up the test tube. Warn your students that if there is any chance that the hot, molten sodium chlorate will rise far enough to come in contact with the glass-wool plug, they should remove the flame immediately and let the test tube cool a bit before reheating. The test tube must be heated until this foam disappears and the liquid in the bottom of the test tube solidifies, and then thoroughly heated for another 10 minutes. The burners can be moved up and down the length of the test tube to heat all the material thoroughly.

When several test tubes of gas have been collected, the stopper can be removed from the generating tube and the oxygen allowed to escape into the air. The gas is tested with a glowing splint, which bursts into flame. The sodium chloride forms a hard, solid mass that may be difficult to remove from the test tube. Usually a scoopula is sufficient to pry the solid loose. A heavier implement, such as a screwdriver, although possibly more effective, is more likely to break the test tube. Grinding up the remaining solid makes it easier to dissolve the sample.

As preparation for the concept of constant proportions (Sec. 6.5) and the possibility of incomplete or slow reactions (Sec. 6.8), you might ask in the post-lab, "When is the reaction complete?" "How can you be sure?" "What if you mass the test tube and contents, heat another 10 minutes, mass again, and find a loss in mass?"

Sample Data

The relative solubility of the original substance and the product was determined at room temperature in 5 cm³ of water. The material was added 1 g at a time to the water in a test tube and shaken vigorously.

	1 g	2 g	3 g	4 g	5 g
Initial solid (sodium chlorate)	Soluble	Soluble	Soluble	Soluble	Excess solid
Final solid	Soluble	Soluble*	Excess solid	—	—

*Actually, 2 g of pure sodium chloride will not quite all dissolve in 5 cm³ of water at room temperature, but it is difficult to convert all the chlorate to chloride, and so the resulting mixture will have a solubility greater than that of pure sodium chloride.

Answers to Questions

When sodium chlorate in a test tube is heated by using two alcohol burners, the solid melts. It then begins to bubble and foam, giving off a gas, and a solid remains at the end.

The remaining solid has a higher melting point than the original substance, because the sodium chlorate melts easily when heated by the alcohol burners, but the residue remains solid even with prolonged heating.

The solubility of the remaining substance is less than that of sodium chlorate.

One can conclude from this experiment that sodium chlorate decomposes into oxygen and a solid that is different from sodium chlorate. Mixing oxygen and the final solid will not result in sodium chlorate.

Apparatus and Materials

Pegboard
Balance
3 Small test tubes (20 × 150 mm)
2 Small clamps
2 Large clamps
2 Alcohol burners
Graduated cylinder (10 cm³)
Sodium chlorate (10 g)
Wood splints
No. 2 one-hole stopper
Right-angle glass bend
Rubber tubing
Glass wool
Bucket
Scoopula
Water
Matches
Paper towels
Mortar and pestle
Safety glasses

EXPERIMENT: DECOMPOSITION OF WATER   6.2

This experiment, like the previous one, shows the decomposition of a pure substance into two new substances, but electrolysis is used instead of heating to decompose the substance. These new substances possess quite different characteristic properties from each other and from the original substance. In addition, the ratio of the masses of the two gases formed is found to be close to $1/8$. The mass ratio is an example of constant proportions, which will be discussed in Sec. 6.5.

Like heat in the previous experiment, electricity is used here only as a tool that allows us to find out more about matter. Do not attempt to explain the mechanism of the reaction.

124 Compounds

At the end of the experiment, the word *compound* is defined and used for the first time. From this point on, the word will be used often.

The Experiment

The saturated sodium carbonate solution used in this experiment can be prepared by mixing 150 g of the monohydrate $Na_2CO_3 \cdot H_2O$ (Arm and Hammer concentrated washing soda) with 400 cm^3 of water. You will notice that the solution is saturated. The apparatus is set up as in Fig. 6.3. Several practical difficulties can be avoided if you demonstrate to the class the following procedure. First, fill a test tube with water and cover the end with a piece of paper or rubber, making sure there are no bubbles in the tube. Then invert the tube, and place it in a 250-ml beaker less than half filled with water. Insert the electrodes into the mouth of the tubes, and clamp the tubes and electrodes into place.

Too much water in the beaker at the start of the experiment will result in very slow evolution of gas, and as the water is displaced from the test tubes, the beaker may overflow. After the apparatus is set up, the water level can be lowered by using a syringe or a roasting baster.

Little or no gas will be visible until the sodium carbonate solution is added. Adding from 10 cm^3 to 30 cm^3 of the solution to 100 cm^3 of water gives a reasonable rate of electrolysis. This rate depends on the voltage, the concentration of electrolyte, the surface area of the electrodes, and the distance through the electrolyte between the electrodes. One can collect a test tube of hydrogen in 15 minutes with a 30 cm^3 of solution, a 6-volt D.C. source, and the electrodes one inch apart and *below* the level of the test-tube lip. (See Fig. 6.3. If the test tubes are too high above the electrodes, some gas may rise outside of the test tubes.)

After testing the gases, the volume of gas produced can be measured by pouring water from a graduated cylinder up to the mark made on the test tubes.

When collecting both hydrogen and oxygen in the same test tube, use a large-size test tube that will accommodate both electrodes.

Caution your students not to let the electrodes touch each other inside the tube while generating gas. If a spark occurs in the gas, a small explosion will occur and spatter the solution. The 2-to-1 mixture of the gases will give a loud "bang" when ignited. It is quite harmless with an open test tube.

Sample Data

	Volume of Hydrogen (cm^3)	Volume of Oxygen (cm^3)	Volume of Hydrogen / Volume of Oxygen	Mass of Hydrogen / Mass of Oxygen
Trial 1	36	18	2.0	0.13
Trial 2	38	19	2.0	0.13
Trial 3	34	16	2.1	0.14
Trial 4	33	15	2.2	0.14

Compounds 125

Answers to Questions

Nothing happens until sodium carbonate solution is added and stirred. The amount of solution added to the water does not affect the volume ratio of hydrogen to oxygen.

When the two gases are collected in one test tube and ignited, there is a small explosion, from which it can be concluded that the gases readily combine. Since the test tube is wet before ignition, it is not possible to conclude that water is formed as a result of the reaction.

Apparatus and Materials

Pegboard	Sodium carbonate (10–30 cm^3)
2 Small test tubes	2 No. 2 solid stoppers
Large test tube	Stirring rod
2 Small clamps	Marking pencil
Large clamp	Water
Beaker (250 cm^3)	Paper towels
Electrodes and leads	Splints
Battery (6-volt) or other 6-volt D.C. source	Matches
	Safety glasses
Graduated cylinder (50 cm^3)	

THE SYNTHESIS OF WATER 6.3

Because they decomposed water in Expt. 6.2 and "know" that water is H$_2$O, students are all too ready to accept the statement that when you synthesize water, 2 volumes of hydrogen combine with 1 volume of oxygen. Thus they may skim over the details of this section. Discuss thoroughly the details of the experiment in Fig. 6.4 and the related data in Table 6.1. The point to emphasize is that the relative amounts of hydrogen and oxygen before reaction do not determine the ratios of these substances that actually combine.

Make sure that your students realize that the water that rises in the tubes in Fig. 6.4 comes from a reservoir (the long tray, not shown in the figure) and is not the water produced from the combination of the hydrogen and oxygen. Problems 7, 8, and 18 will help to illustrate this point.

This experiment is dangerous. Do not demonstrate.

EXPERIMENT: SYNTHESIS OF ZINC CHLORIDE 6.4

Different groups are given different amounts of zinc to react with the same amount of hydrochloric acid. The results of the entire class are then compared by means of a histogram. Only by varying the mass of zinc over a wide enough range can we establish that the ratio of the mass of the zinc reacting to the mass of the product is constant regardless of whether there was an excess of hydrochloric acid or an excess of zinc.

It is particularly important that you take time for a thorough post-lab class discussion after your students have completed the experiment. The data from all groups should be tabulated on the chalkboard in the manner shown under Sample Data, and a histogram should be made of the results.

This experiment will require two periods. Before beginning, make sure that your students understand the reason for varying the amount of zinc while keeping the amount of acid constant. Allow plenty of time at the start so that the initial massing can be done carefully. The reaction should be started and then set aside till the next day, when the reaction is complete and the solution can be evaporated to dryness. If you have more than one class, have a few extra evaporating dishes handy in case some of the lab groups do not finish and need to store their partially dried zinc chloride overnight.

The Experiment

Individual groups should be given one to ten pieces of zinc and 10 cm^3 of hydrochloric acid. Mossy zinc is too impure and should not be used.

The acid should be 6N, prepared by mixing equal volumes of concentrated hydrochloric acid and water. When diluting concentrated acid, remember that the acid must *always* be poured into water, never the other way around.

The experiment has been designed in such a way that 1, 2, 3, and 4 pieces of zinc are completely used up in the reaction, leaving an excess of acid. With 5 or more pieces there is an excess of zinc. The strength and the amount of acid and the amount of zinc are fairly critical if the break-even point is to be between 4 and 5 pieces.

When the liquid is poured into the evaporating dish, a wood splint can be used to prevent the remaining pieces of zinc from falling out of the test tube. Tiny specks of material that flow out of the test tube with the solution are of negligible mass and will not appreciably affect the results. While the solution is evaporating, the unreacted pieces of zinc can be massed. They should first be dried carefully on a paper towel. A few minutes between drying and massing will allow any remaining moisture to evaporate. Small flecks of material remaining on the paper towel are of negligible mass.

The evaporating dish containing the solution can be heated with an alcohol burner. The amount of material lost through spattering is small. After about 20 minutes, the zinc chloride residue will form a puffy white crust. Heating should continue until the solid melts to be sure that the water has been driven off. The zinc chloride formed in the experiment is a hydrate. When it is dried, it becomes the anhydrous form, which has a low melting point.

The evaporating dish and contents should be massed just as soon as the dish is cool enough to touch (after about 5 minutes), because the zinc chloride will pick up moisture rapidly from the air. If for some reason the residue cannot be massed immediately after heating, it should be reheated before massing to remove the water that has been absorbed.

Sample Data

Pieces of Zinc Used	Mass of Zinc Used (g)	Mass of Zinc Remaining (g)	Mass of Zinc Reacted (g)	Mass of Product (g)	Mass Zinc Reacted / Mass Product
1	0.46	0	0.46	0.94	0.49
2	0.92	0	0.92	1.84	0.50
2	0.94	0	0.94	1.96	0.48
3	1.36	0	1.36	2.78	0.49
3	1.36	0	1.36	2.78	0.49
4	1.79	0	1.79	3.80	0.47
4	1.82	0	1.82	3.75	0.49
5	2.30	0.07	2.23	4.78	0.47
5	2.32	0.44	1.88	3.86	0.49
6	2.65	0.46	2.19	4.67	0.47
6	2.65	0.51	2.14	4.39	0.49
7	3.20	0.94	2.26	4.81	0.47
8	3.63	1.38	2.25	4.72	0.48
9	4.07	1.84	2.23	4.78	0.47
10	4.46	2.35	2.11	4.53	0.47

Answers to Questions

The gas given off in the reaction is hydrogen.

The water used to wash out the test tube is added to the solution in the evaporating dish because it contains some of the dissolved product (zinc chloride).

The ratio of the mass of zinc reacted to the mass of product formed is close to 0.48, as shown by the histogram in Fig. I.

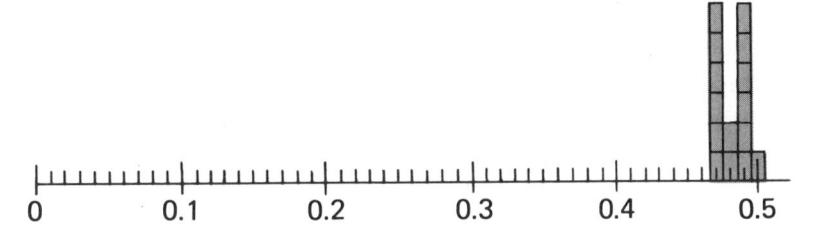

Mass of zinc reacted / Mass of product

Figure I

When 1 to 4 pieces of zinc are used, there is obviously an excess of hydrochloric acid; whereas with 5 to 10 pieces of zinc there is an excess of zinc. This excess of one component does not affect the final ratio.

If the zinc chloride were not completely dry when massed, the mass of the product would be larger, and the final ratio would therefore be smaller.

The purpose of probs. 11 and 12 is to help students to understand the reasoning that leads to constant combining ratios.

Apparatus and Materials

Pegboard
Balance
Large test tube (25 × 150 mm)
Large clamp
Graduated cylinder (10 cm³)
Beaker (250 cm³)
Evaporating dish (000)
2 Alcohol burners
Burner stand

Zinc pieces (approximately 0.45 g each; average of 5 pieces per lab team)
6N hydrochloric acid (10 cm³)
Water
Paper towel
Matches
Wood splint
Safety glasses

6.5 THE LAW OF CONSTANT PROPORTIONS

Treat this section thoroughly. This is a good opportunity to go back to Secs. 2.14 and 2.15 and review what we mean by a law of nature. A good understanding of constant proportions will be helpful in the discussion of multiple proportions in Chap. 8.

All the available data did not clearly support either Berthollet or Proust, because the techniques of massing and separation were not sufficiently sophisticated. However, before Proust published his 1799 paper, he had done a great deal of quantitative work on the black oxide of copper, and this work formed the basis for his generalization.

It is difficult to generate excitement about a controversy that occurred in science over 150 years ago, and students generally see no connection between it and current science. However, in a modern geologic controversy, for example, the balance swung only recently in the direction of the "continental drift theory"—the idea that the continents have drifted thousands of miles apart over long spans of time, instead of having been formed where they are now.

6.6 EXPERIMENT: A REACTION WITH COPPER

When copper dust in a crucible is heated in air, the reaction proceeds very slowly. The student can see that something is happening as the material changes from the characteristic color of copper to the dull black of copper oxide. The reaction may appear to be complete after 2 min of heating. Since this experiment is designed to give a mixture of copper and copper oxide needed in the following experiments, the copper should not be broken up or stirred until the end of the experiment.

The gain in mass of the contents of the crucible is further evidence that a reaction has taken place. Further heating shows an additional gain in mass, and there is no guarantee that the reaction is complete after 17 min of heating. Upon breaking up the black solid after 17 min of heating, a few specks of shiny copper may be seen, which suggests that the reaction may be nearly complete. The large amount of unreacted copper obtained at the conclusion of the next experiment will come as a surprise. The reaction was far from complete!

The Experiment

The success of this experiment and of the one that follows depends on using fine copper dust.

Figure II shows the clay pipe-stem triangle that holds the crucible containing copper. Two of the twisted pairs of wires are squeezed together almost parallel, and the ends of the individual wires in each pair are bent (one up and one down) as illustrated. The two wires bent upward can be pushed through holes in the pegboard, hooking the triangle firmly in place.

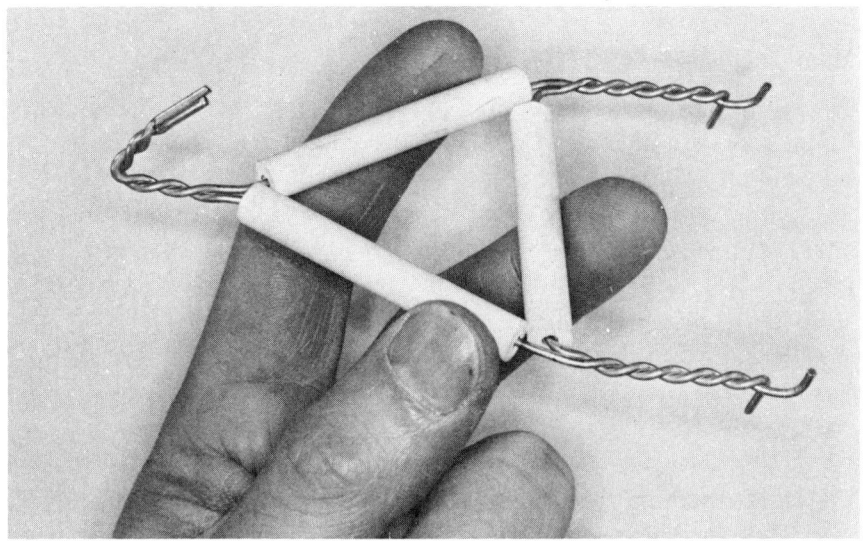

Figure II

In a minute or so after heating has begun, the copper comes up to ignition temperature, and the reaction can be seen to spread across its surface. At this point the reaction is nowhere near complete, although your students may think otherwise.

After 2 min of heating, 1.00 g of copper will gain about 0.01 g. After 17 min the total gain in mass will be about 0.03 g to 0.04 g.

Answers to Questions

The copper turns black in about 1 to 2 min, and the reaction appears to be complete.

The crucible containing the copper has gained mass. This is further evidence that the copper combined with something in the air.

More heating may or may not result in a further increase in mass. After an additional 15 min of heating, the mass increased still more.

Only a few specks of shiny copper can be seen upon breaking up the contents of the crucible after 17 min of heating. From this it can be estimated that almost all the copper reacted.

Apparatus and Materials

Pegboard
Balance
Porcelain crucible (00)
Clay triangle
Alcohol burner and fuel
Copper dust (1 g)
Test tube
Test-tube rack
Scoopula
Safety glasses
Matches

6.7 EXPERIMENT: THE SEPARATION OF COPPER OXIDE FROM A MIXTURE OF COPPER OXIDE AND COPPER

Be sure to demonstrate that a little copper dust in a test tube does not dissolve in hydrochloric acid.

In this experiment students can readily see that the reaction in Expt. 6.6 was incomplete; when the hydrochloric acid dissolves the copper oxide, a large residue of uncombined copper is observed.

The Experiment

There are no problems in the performance of this short experiment. Emphasize care in the use of the hydrochloric acid; it is irritating to the skin and can badly damage some types of cloth.

The copper can be washed by adding 5-10 cm^3 of water and shaking and stirring. The first washing may show a small amount of white precipitate, which disappears after a second washing. The test tube containing the dark green solution of copper oxide in acid should be stoppered with a rubber stopper, labeled, and stored for use in Expt. 7.1.

Answers to Questions

The solid remaining in the test tube looks like copper.

The substance left in the crucible at the end of the last experiment evidently was a mixture.

Apparatus and Materials

2 Test tubes (20 × 150)
Test-tube rack
Graduated cylinder (10 cm^3)
Stirring rod
Test tube containing black solid from Expt. 6.6
5 cm^3 6N hydrochloric acid (same concentration as used in Expt. 6.4)

6.8 COMPLETE AND INCOMPLETE REACTIONS

Treat this section as a reading assignment. The students will understand some of the causes of the controversies referred to in Sec. 6.5, but not all. Other factors entering into the discussion appear in Chap. 8.

CHAPTER 6—ANSWERS TO PROBLEMS

Sec. Sec.	Easy	Medium	Hard	Class Discussion	Home or Lab
1	1†		16, 17	17	1
2	3†, 4†	2, 5		5	
3		6†, 7, 9†	8, 18	8	
4	11†, 12, 13†	10, 19, 24		12, 19, 24	
5	15†	14, 22	20	22	
6-7			23		
8		21			

1† Suppose that in decomposing sodium chlorate, you did not heat the sodium chlorate long enough to decompose all of it.
   a) What substances would be left in the test tube?
   b) How would you separate them?

   a) Sodium chlorate and sodium chloride
   b) By fractional crystallization. (Interested students might try this.)

2 In decomposing water, suppose you had filled many test tubes with gas, adding water to the beaker as it disappeared but never adding more sodium carbonate. You would always have found the ratio of hydrogen to oxygen produced to be constant. What does this tell you about the source of the gases? About the sodium carbonate?

The gases come from the water and not from the sodium carbonate, which is not used up.

3† a) What is the total mass of oxygen and hydrogen that can be produced by the decomposition of 180 g of water by electrolysis?
   b) If all the hydrogen produced were burned in the air to form water, what mass of water would result?

   a) 180 g
   b) 180 g

4† Two test tubes contain equal volumes of gas at atmospheric pressure. If one contains oxygen and the other helium, is the mass of gas in both tubes the same?

No.

132 Compounds

5 From your data in Expt. 6.2, calculate the ratio of the mass of oxygen produced to the mass of hydrogen produced. How does your ratio compare with those of the other members of your class?

Suppose that a student collected 23.3 cm³ of hydrogen and 12.3 cm³ of oxygen. Then the mass of oxygen and hydrogen produced is

Mass of oxygen = 12.3 cm³ × 1.3 × 10⁻³ g/cm³ = 1.6 × 10⁻² g.
Mass of hydrogen = 23.3 cm³ × 8.4 × 10⁻⁵ g/cm³ = 1.96 × 10⁻³ g.

Thus the mass ratio of oxygen to hydrogen is

$$\frac{1.6 \times 10^{-2} \text{ g}}{1.96 \times 10^{-3} \text{ g}} = 8.2$$

Most other members of the class are likely to have ratios of 8.2 ± 0.5.

6 † If 18 g of water is decomposed into hydrogen and oxygen by electrolysis, 16 g of oxygen and 2 g of hydrogen are produced. Using the table of densities in Chap. 3 (page 43), find (a) the volume of water decomposed and (b) the volume of hydrogen produced.

a) 18 cm³
b) 2.4 × 10⁴ cm³

7 You mix 100 cm³ of oxygen with 200 cm³ of hydrogen. The volumes of both gases are measured at atmospheric pressure and at room temperature.

a) Calculate the mass of oxygen used and the mass of hydrogen used.
b) If you ignited the mixture, what mass of water would result from the reaction?
c) What volume of water would be produced?

a) Mass of oxygen = 100 cm³ × 1.3 × 10⁻³ g/cm³ = 0.13 g
    Mass of hydrogen = 200 cm³ × 8.4 × 10⁻⁵ g/cm³ = 0.017 g
b) Mass of water = 0.13 g + 0.017 g = 0.15 g
c) Volume of water = 0.15 g/1.00 g/cm³ = 0.15 cm³

8 If in Question 7 you had used 100 cm³ of oxygen but only 50 cm³ of hydrogen, what mass of water would have resulted?

The masses of oxygen and hydrogen can be calculated as in the preceding problem, since we know that 25 cm³ of oxygen react with 50 cm³ of hydrogen.

Mass of oxygen = 25 cm³ × 1.3 × 10⁻³ g/cm³ = 0.032 g
Mass of hydrogen = 50 cm³ × 8.4 × 10⁻⁵ g/cm³ = 0.0042 g
Mass of water = 0.032 + 0.0042 g = 0.036 g

More simply, however, if ¼ the amount of hydrogen was used, ¼ the amount of water must have been produced.

$$\text{Mass of water} = \frac{1}{4} \times 0.15 \text{ g} = \underline{0.037 \text{ g}}$$

9† Three tubes are filled with a mixture of hydrogen and oxygen in a manner similar to that used in Fig. 6.4. If the three tubes contain the following volumes of hydrogen and oxygen, what is the volume of the *unreacted* gas remaining in each tube after they are ignited?

Tube	Volume of oxygen (cm³)	Volume of hydrogen (cm³)
I	25	75
II	50	50
III	25	50

Tube I: $\underline{25 \text{ cm}^3}$
Tube II: $\underline{25 \text{ cm}^3}$
Tube III: $\underline{0 \text{ cm}^3}$

10 Suppose that in the synthesis of zinc chloride you dissolved 5 g of zinc.
   a) How much product would you get?
   b) What would be the ratio of zinc to the product?
   c) What would the ratio be if you dissolved 50 g of zinc?

The experiment yields (Mass of zinc)/(mass of product) = 0.48. Using this value:
a) Mass of product = 5.0 g/0.48 = $\underline{10.4 \text{ g}}$
b) $\underline{0.48}$
c) The ratio (Mass of zinc)/(mass of product) remains the same, no matter how much zinc is dissolved.

11† In a certain package of seed corn the number of red seeds was 36, and the number of yellow seeds was 24. In a second package the number of red seeds was 51, and the number of yellow seeds was 34.
   a) What is the ratio of the number of red seeds to the number of yellow seeds in each package?
   b) What is the ratio of the number of red seeds to the total number of seeds in each package?

	First package	Second package
a)	$\underline{1.5}$	$\underline{1.5}$
b)	$\underline{0.6}$	$\underline{0.6}$

134   Compounds

12 Suppose that there are 10 boys and 15 girls in a class. (a) What is the ratio of boys to girls in the class? (b) What is the ratio of boys to total number of students in the class? (c) What would be the ratio of boys to girls if the class were three times larger but the ratio of boys to total number of students was the same?

a) $\dfrac{\text{No. of boys}}{\text{No. of girls}} = \dfrac{10}{15} = \dfrac{2}{3} = \underline{0.67}$

b) $\dfrac{\text{No. of boys}}{\text{Total no. of students}} = \dfrac{10}{25} = \dfrac{2}{5} = \underline{0.40}$

c) As long as the ratio (No. of boys)/(No. of students) remains the same, the ratio (No. of boys)/(No. of girls) also remains the same.

Since we shall use ratios of this sort in many later calculations, it is worthwhile to make sure the students understand this idea now by giving them a few more examples. For instance, if 3 parts of A combine with 7 parts of B to give 10 parts of C, the ratio of $A/B = 0.43$, $A/C = 0.3$, or $B/C = 0.7$ will always remain the same for the masses of A, B, and C regardless of the total mass of C formed.

13† When various amounts of zinc react with hydrochloric acid, zinc chloride and hydrogen are produced. Which of the following ratios of masses between the various products and reacting substances are constant regardless of the amounts of zinc and acid mixed together?

(1) $\dfrac{\text{Zinc added}}{\text{Zinc chloride produced}}$

(2) $\dfrac{\text{Zinc used up}}{\text{Hydrochloric acid used up}}$

(3) $\dfrac{\text{Zinc used up}}{\text{Hydrogen produced}}$

(4) $\dfrac{\text{Zinc used up}}{\text{Zinc chloride produced}}$

(5) $\dfrac{\text{Zinc added}}{\text{Hydrochloric acid added}}$

(6) $\dfrac{\text{Zinc chloride produced}}{\text{Zinc used up}}$

(7) $\dfrac{\text{Hydrochloric acid used up}}{\text{Hydrogen produced}}$

Ratios: $\underline{2, 3, 4, 6, 7}$

14 If you make a solution of salt and water, over what range of values can you vary the mass ratio of salt to water at a given temperature?

The mass ratio can be near zero when there is very little salt, and it can become larger and larger until the ratio is limited by the solubility of salt in water. The ratio at this point is 0.36 at 20°C for sodium chloride and cannot be increased except by raising the temperature of the solution. Below this point the mass ratio is not fixed.

15† a) Do you think that gasoline is a single compound? See Table 5.1.
b) Would you expect gasoline from different pumps to be the same?

16. a) How would you have changed the procedure you followed in Expt. 6.1 if you had wanted to measure the density of the gas?
b) Would you have had to decompose all the sodium chlorate in order to determine the density of the gas?

a) The mass of the test tube and the material in it must be measured both before and after heating. All the gas produced must be collected in a large bottle and the volume measured.
b) The density of the gas can be determined without decomposing all the sodium chlorate; but the larger the change in the mass, the more accurate will be the determination of the density.

17. Ten grams of sodium chlorate was strongly heated. After the heating it was found that 5 cm³ of water at 100°C was required to completely dissolve a 2-g sample of the residue.
a) What was the solubility of the residue in grams per 100 cm³ of water at 100°C?
b) Was any sodium chlorate left in the residue?

a) Since 2 g of the sample just dissolves in 5 cm³ of water at 100°C, 100 cm³ of water will dissolve $2 \text{ g} \times \frac{100 \text{ cm}^3}{5 \text{ cm}^3} = 40$ g.

b) The solubility graph in Fig. 6.2 shows that sodium chloride has a solubility of 40 g/100 cm³ of water at 100°C. If there were any appreciable amount of sodium chlorate in the residue, the solubility would be more than 40 g/100 cm³ of water at 100°C; the residue must be nearly all sodium chloride.

18. Suppose that the apparatus used in Expt. 6.2, Decomposition of Water, contained 100 cm³ of water. Using this apparatus, a student collected 57 cm³ of hydrogen and 28 cm³ of oxygen. What fraction of the total volume of water was decomposed?

To find the volume of the water that was decomposed, let us first find its mass. The mass of the water equals the mass of hydrogen and oxygen obtained and can be calculated as follows:

Mass of hydrogen = 57 cm³ × 8.4 × 10⁻³ g/cm³ = 4.8 × 10⁻³ g
Mass of oxygen = 28 cm³ × 1.3 × 10⁻³ g/cm³ = 36 × 10⁻³ g

The total mass of water decomposed is 5 × 10⁻³ g + 36 × 10⁻³ g = 41 × 10⁻³ g, and therefore, since the density of water is 1.0 g/cm³, the total volume of water is 41 × 10⁻³ cm³.

The fraction of the total volume of water decomposed is

$$\frac{41 \times 10^{-3} \text{ cm}^3}{100 \text{ cm}^3} = 41 \times 10^{-5}$$

136 Compounds

19 About a gram of salt is placed in a test tube half full of water and shaken; about a gram of citric acid is placed in a test tube filled with alcohol and shaken; some magnesium carbonate is dropped into half a test tube of dilute sulfuric acid; hydrochloric acid is poured into a dish containing magnesium. In each case, the solid disappears, and we say that it "dissolved." However, it is evident that two different kinds of dissolving have occurred. Divide the experiments into two classes. What did you observe that led you to divide them this way? What do you think you will observe in each case if you evaporate the solution to dryness?

Acids act on metals with the evolution of a gas, but this does not occur with salt and water or with citric acid and alcohol. Evaporation of the liquids will result in the recovery of the salt and the citric acid, but the solid residues left after evaporating the acids in which the metals dissolved are quite different from the original metals. Evidently there are two quite different processes involved.

Class I	Class II
Salt in water	Magnesium carbonate in sulfuric acid
Citric acid in alcohol	Magnesium in hydrochloric acid

The use of the word "dissolve" is often restricted to the cases in which the original solid can be recovered intact.

20 A mass of 5.00 g of oxygen combines with 37.2 g of uranium to form uranium oxide.
   a) How many grams of the oxide are formed?
   b) What is the ratio of uranium to oxygen in this compound?
   c) How much oxygen is needed to completely oxidize 100 g of uranium?

   a) From the law of conservation of mass, the mass of the oxide must be 5.00 g + 37.2 g = 42.2 g.

   b) $\frac{\text{Mass of uranium}}{\text{Mass of oxygen}} = \frac{37.2 \text{ g}}{5.00 \text{ g}} = \underline{7.44}$

   c) Mass of oxygen = 100 g/7.44 = $\underline{13.4 \text{ g}}$

21 Discuss the process of boiling an egg in terms of complete and incomplete reactions. What may affect the time required for a complete reaction?

If the egg is heated only a short while in boiling water, the white will remain partially liquid. When the egg has been boiled long enough to solidify both the white and the yolk, the reactions are complete, and continued heating will cause no further change.

22 What evidence do you have for the following statements?
a) Zinc chloride is a pure substance and not a mixture.
b) Sodium chlorate is a pure substance and not a mixture of sodium chloride and oxygen.
c) Water is a pure substance.

a) The ratio (Mass of zinc reacted)/(Mass of zinc chloride) is constant, regardless of how much zinc is added to hydrochloric acid.
b) Mixing sodium chloride and oxygen (or air, which contains oxygen) does not produce a substance with the properties of sodium chlorate. Also, sodium chlorate melts at a lower temperature than sodium chloride.
c) Water has a definite melting point. When hydrogen and oxygen are mixed in different ratios and ignited, they combine in a fixed ratio and form water.

23 A student heated a crucible containing 2.15 g of powdered copper until the mass of the contents of the crucible became 2.42 g. The black solid in the crucible was placed in hydrochloric acid, and 1.08 g of copper remained undissolved in the acid. What is the reacting ratio by mass of copper with oxygen?

2.15 g − 1.08 g = 1.07 g of copper that reacted.
2.42 g − 2.15 g = 0.27 of oxygen that combined with copper.

Therefore, the ratio $\dfrac{\text{mass of oxygen reacted}}{\text{mass of copper reacted}} = \dfrac{0.27 \text{ g}}{1.07 \text{ g}} = \underline{0.25}$

Or the ratio $\dfrac{\text{mass of copper reacted}}{\text{mass of oxygen reacted}} = \dfrac{1.07 \text{ g}}{0.27 \text{ g}} = \underline{4.0}$

24 Hydrochloric acid is a solution of the compound hydrogen chloride in water. Suppose there were two bottles containing hydrochloric acid in your lab. How would you determine whether the concentration of hydrogen chloride in each bottle is the same?

You can place equal masses of zinc in two test tubes. To one test tube add acid from one bottle, a little at a time–say 1 cm³, until all the zinc has just dissolved. You can do the same with the other test tube and the other bottle of acid. If the concentrations are the same, the same volume of acid will be required to dissolve each sample of zinc. If not, the smaller volume of acid required would be the more concentrated.
　　You could also make a careful determination of the density of the liquid in each bottle. More concentrated hydrochloric acid is denser than less concentrated hydrochloric acid.

Elements

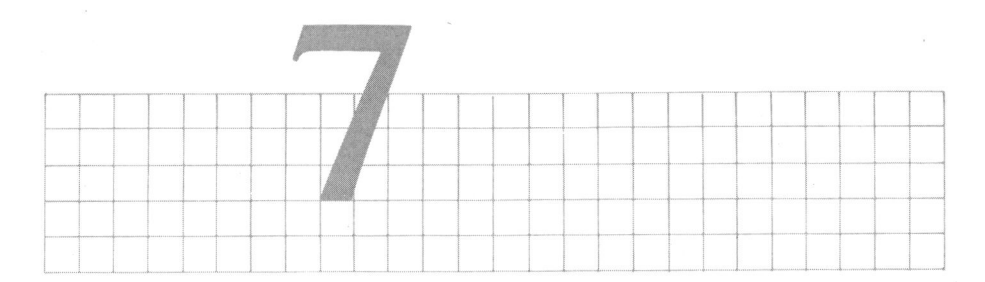

7

Overview of the Chapter

In Chap. 5 we considered the differences between mixtures and pure substances. In Chap. 6 we identified compounds as pure substances that can be separated by suitable means into simpler pure substances; we found that compounds can be synthesized from simpler pure substances, but only in certain fixed proportions. We now turn our attention to those pure substances that resist being broken up into simpler pure substances by the same means that break up most pure substances. These we identify as elements. It may be worthwhile to read parts of Secs. 7.2 and 7.3 in class, but in any case be sure to spend enough time on these sections to make students well aware of their contents.

We now turn to spectral analysis, which fulfills two functions. First, it permits the identification of minute amounts of substances, and second, it enables us to identify elements which make up a compound. (Actually, very small amounts of the compounds which are put in the flame break up into elements, but this does not concern us here.) Precise spectral analysis calls for refined apparatus and accurate measurements. Here we are satisfied with showing the basic phenomenon of flame colors and pointing out how one proceeds from them to the analysis of the line spectra of the elements.

With the concept of *element*, we reach one of the high points of the course. It is now possible to visualize the myriads of materials in the physical world as being made up of a relatively small number of basic substances. However, the picture of matter is not complete. Radioactivity is now introduced for the double purpose of completing our description of elements and of suggesting an atomic picture of matter. The black dots on a photographic emulsion exposed to polonium, the random individual clicks from a Geiger counter, and the separate tracks diverging randomly, in both direction and time, from a radioactive source in a cloud chamber provide evidence for the discrete nature of matter and lead into the next chapter, in which an atomic model of matter is carefully developed. In Chap. 10, radioactive disintegration will be used again, this time to determine the mass and size of atoms.

140 Elements

The historical route by which the atomic model of matter was developed is long, varied, and indirect. For this reason the radioactivity phenomena are used as a direct means of *suggesting* the discreteness of matter. The validity of the atomic model does not come from what is seen in this chapter but later where its usefulness in correlating many of the properties of matter already encountered is discussed. This will be done in Chap. 8.

Radioactivity is a subject that appeals to students, and they will want to know more about it. Nevertheless, topics such as different kinds of radiation and their properties, isotopes, and complete radioactive decay series are all extraneous to the purpose of this chapter and should be left out. A discussion of them soon leads into a discussion of electrons, protons, neutrons, and isotopes for which there is no experimental evidence in this course. It is best to refer interested students to private study. A good book for this purpose is *The Restless Atom*, by Alfred Romer (Science Study Series, Doubleday & Company, Garden City, New York, 1960).

Radioactivity shows students that not all elements have the permanence suggested by their definition as pure substances that cannot be broken up. Students can see in the film loops or demonstrations that some elements (radioactive ones) "expose" photographic film and affect a Geiger counter. They learn from spectrographic evidence that such elements change into other elements by themselves at a rate that cannot be controlled by any of the methods they have used to control reactions in the laboratory.

The suggested schedule for this chapter is:

Sections 1-6 (3 expts., probs. 1-10; 18-20)   7 periods
Sections 7-9 (2 film loops; probs. 11-17; 21-24)   4 periods
Total   11 periods

7.1 EXPERIMENT: ZINC AND COPPER

In Expt. 6.6, the students heated powdered copper and observed an increase in mass and a change in appearance of the copper as it changed to a black solid. In Expt. 6.7, the students added hydrochloric acid to the material (copper oxide and copper) produced in Expt. 6.6, and the copper oxide dissolved, leaving a large amount of unreacted copper. Now we wish to examine the solution formed in Expt. 6.6 to find out whether the copper in the copper oxide can be recovered.

The Experiment

When the quantities of materials called for in Expts. 6.6 and 6.7 have been used, a single piece of zinc (like that used in Expt. 6.4) will be more than sufficient to displace all the copper in the solution.

Before they add the zinc, see that the students, as directed, have placed the test tubes containing the solution in a beaker of water.

Sometimes the copper sticks to the surface of the zinc, which will prevent further reaction. Students should use their stirring rods to shake the copper off the surface of the piece of zinc. After the zinc has precipitated all of the copper, it will begin to react with the hydrochloric acid to form hydrogen gas.

After using the stirring rods the students should immediately dip them into the water in the beaker to dilute the acid. This will prevent acid from dripping onto clothing, books, or skin.

Answers to Questions

The first two questions set the stage for the experiment and cannot be answered until the experiment is complete.

The solid appears to be copper.

Evidently the copper that reacted with oxygen in Expt. 6.6 was dissolved by the acid in Expt. 6.7 and, although it could not be recognized directly, it was a part of the solution. Since it could be recovered, it had not disappeared forever.

Apparatus and Materials

Solution of black solid in
    hydrochloric acid from Expt. 6.7
Zinc pieces (like that used in Expt. 6.4)
250-ml beaker of cool water

Water
Paper towel
Stirring rod
Scoopula

ELEMENTS 7.2

The procedure that we have been following is to break matter down into simpler forms; each new substance we get has specific properties by which we can identify it. When we use a new technique, we evaluate it by its ability to yield more and more substances. Fractional distillation, fractional crystallization, chromatography, electrolysis, and heating have been employed, and characteristic properties have been clues to the use of these methods of separation. For example, a difference in density can be used to separate two solids, as can a difference in solubility. A difference in boiling point enables us to separate liquids.

We have seen examples of pure substances that can be broken down into two or more pure substances. When they do break down, we find by carefully measuring the mass ratio before and after decomposition that they decompose according to a law of constant proportions. These original substances are called compounds.

The separation of a mixture or the decomposition of a compound may produce one or more compounds that also can be decomposed. But, many times, one or more pure substances are produced that cannot be broken down by any method used. We have found that there are over 100 such substances or elements that make up all the known substances on the earth.

TWO SPECIAL CASES: LIME AND OXYMURIATIC ACID 7.3

Before 1808, using the tools available, investigators were unable to decompose lime into any further pure substances. In this sense, lime remained an element until the time of Humphry Davy, who was able to produce a new element, calcium, by using a new tool, electrolysis.

Conversely, Humphry Davy demonstrated that a supposed compound, oxymuriatic acid (chlorine), was *not* further decomposable with a given set of tools and was, therefore, definable as an element.

The operational nature of the reasoning involved in these two cases is the important point to be made by these historical examples.

7.4 EXPERIMENT: FLAME TESTS OF SOME ELEMENTS

In this experiment students see the characteristic colors that some elements impart to a flame. More than one compound of the same element is used to demonstrate that the color comes from the common element and not from the particular compound. The other elements in the compounds do not give any visible light when heated in an alcohol flame. All elements will show a spectrum in a hot enough flame.

It is difficult, by the flame test, to distinguish between the elements calcium, lithium, or strontium.

The spectroscope in Expt. 7.5 can be used at that time to spread the colors out into the bright line spectra of the elements in the sample. A comparison of the spectra of the unknown sample and those of known samples of calcium, lithium, and strontium will show which elements are present.

The Experiment

The solid salts are held in the flame on a nichrome wire. Cut the wire into 3-inch lengths, make as small a loop as possible at one end of each piece, and insert the other end into the soft wood handle.

Be sure that the same nichrome wire is always used with the same salt. The wire is difficult to clean, and if the same piece is used with more than one salt, the characteristic colors of both salts may be observed simultaneously. Moreover, when the same wire is dipped in different bottles, the salts get mixed. The nichrome wires can be labeled with the name of the salt with which they are to be used. One way to reduce contamination is to give a pair of students in a group one substance (and accompanying nichrome wire) at a time. Do not give them the next substance until they have finished and passed the substance on to the next pair in the group.

The salts may drop off the holder onto the burner. However, if the burner is set at an angle in a clamp on the pegboard, there is little difficulty with contamination of the burner. An alcohol burner is hot enough to produce the characteristic color of each metal, but a Bunsen burner gives brighter colors.

The spectra kit used in this experiment includes compounds of potassium, which although not mentioned in the text, can also be tested in the flame of an alcohol burner.

Sample Data

Sodium compounds. When a compound containing sodium is introduced, a bright yellow flame results.

Copper and copper compounds. These impart a greenish color to the flame. With copper chloride some blue can be seen in addition to the green.

Strontium compounds. Strontium compounds give the flame a scarlet red color that disappears rather quickly.

Lithium compounds. Lithium gives a brilliant carmine color.

Calcium compounds. These give a brick-red color that may be partially obscured if there is much sodium present in the flame.

Potassium compounds. Potassium produces a violet flame. If there are any sodium impurities, or if the flame itself is at all yellow, the violet color is masked. Cobalt glass can be used to filter out the yellow light.

Answers to Questions

Sodium is distinguished by its yellow color. Copper and copper compounds impart a green color to a flame.

It is difficult to distinguish between strontium, calcium, and lithium compounds by the colors they impart to a flame.

Apparatus and Materials

Copper powder
Copper oxide (1 g)
Alcohol burner
Paper towels
Matches
In Spectra Kit:
    Calcium chloride (5 g)
    Copper chloride (5 g)
    Copper nitrate (5 g)
    Lithium chloride (5 g)

Potassium chloride (5 g)
Potassium nitrate (5 g)
Sodium hydrogen
    carbonate (5 g)
Sodium chloride (5 g)
Strontium chloride (5 g)
Strontium nitrate (5 g)
15 nichrome wires
    with handles

EXPERIMENT: SPECTRA OF SOME ELEMENTS 7.5

In this experiment students will use a small grating spectroscope to observe the light emitted from several familiar light sources. A comparison is made between the appearance of a familiar, rainbowlike, continuous spectrum and the kind of line spectra that can be used to identify elements. The spectroscope extends the senses in distinguishing colors.

The Experiment

The spectrum of an incandescent lamp is a continuous gradation from the blue end to the red end of the spectrum. The spectrum of a fluorescent lamp is composed of a continuous spectrum that comes from the white phosphorescent coating, on which appear lines that are caused by the mercury vapor inside the tube. The resulting spectrum looks like that obtained from the incandescent bulb, but with several bright lines superimposed. Students will not be able to identify the line spectrum as that of mercury, but they will have seen a line spectrum for themselves, which will make the discussion of spectra in this chapter more real to them.

If you have only alcohol burners, it is difficult to see the line spectra of salts in a flame. If you have Bunsen burners, a yellow line caused by a sodium flame is clearly visible. A red line and a yellow line are visible for lithium. Other salts produce color for only a few seconds, and the line spectra are very difficult to observe with a Bunsen burner.

The room should be darkened, and the spectroscope should be a few feet from the flame in order to see the spectra. It is best to mount the spectroscope rigidly on a pegboard with two clamps, and then align the flame with the spectroscope before introducing the salts.

A red flare (available at automotive or hardware stores) produces a number of bright lines as well as a continuous spectrum. Do not hold the flare in your hand. Because of fumes, use the flare only under a hood or outdoors. Most flares will continue to burn for 15 to 20 minutes; however, they can be extinguished by putting the burning end of the flare into water.

Students can be asked to draw the spectra they observe. If this is to be done, perhaps a form like that diagrammed below could be distributed. In this way the student would be made aware of the placement of spectral lines relative to each other and also of the region of the total spectrum in which the lines are falling.

Violet	Blue	Green	Yellow	Orange	Red

Answer to Question

The difference between the spectrum of a filament bulb and that of a fluorescent lamp is the bright lines in the latter.

Apparatus and Materials

Grating spectroscope
Incandescent lamp
Fluorescent lamp

7.6 SPECTRAL ANALYSIS

Line spectra are a characteristic property of elements. The method of spectral analysis is introduced in order to have a means of identifying elements in small amounts and of identifying elements in mixtures and compounds. Elements are identified by their spectra in this chapter and in Chap. 10.

RADIOACTIVE ELEMENTS 7.7

Show your classes the *Introductory Physical Science* film loop titled "Radioactive Substances I," in which students witness a simplified version of Becquerel's classic experiment that led to the discovery of radioactivity. Furthermore, the observation of the random counts made by a Geiger counter near a radioactive source is the first of the steps that leads us eventually to a particle model of matter. The experiment is designed in such a way that radioactivity is associated with certain individual elements rather than with the compounds that contain them. As a result, we can now speak of radioactive elements.

RADIOACTIVE DECOMPOSITION 7.8

In this section radioactivity is associated with the decomposition of radioactive elements. Because we have gone to some trouble to establish earlier in the chapter that the substances we call elements are those that cannot be separated into simpler substances by the standard means, the idea of the decomposition of an element may at first be confusing to students. The section treats in depth the profound differences between chemical decomposition and radioactive decomposition and why elements that decay are still called elements. Reading this section with your students will help to ensure that the important points are not missed.

A CLOSER LOOK AT RADIOACTIVITY 7.9

Figure 7.7 shows a photographic plate that was exposed to a very small sample of polonium. It appears that the number of dots produced on the plate depends directly on the exposure time, which is true as long as the exposure time is long enough but not too long compared with the half-life of polonium. The half-life of polonium is 138 days. In the experiment described in the text, the maximum exposure time was 144 hours, or 6 days. Thus in 6 days only 3 percent of the sample has disintegrated, and the activity remains practically unchanged during the experiment. You may want to introduce the idea that the radioactivity of a sample decreases with time, but wait until Chap. 10 for a quantitative presentation of decay curves. The concept of half-life is unnecessary (and confusing to many students) at this point. The dots do not show on the Polaroid film used in the film loop "Radioactive Substances I."

With this section it is important that you show the film loop "Radioactive Substances II." This loop shows the use of the cloud chamber in demonstrating the fog tracks radiating out from a sample of radioactive material; the cloud chamber provides a more graphic demonstration of the emission of particles in all directions from a radioactive source than does a photographic film or a Geiger counter.

If you have cloud chambers available and have access to Dry Ice, you may wish to let students set them up as individual projects. The chamber itself is a plastic box in which a radioactive source can be placed. The source, lead 210, is supplied with the chamber and is on the end of a wire stuck in a cork that fits into a hole in the box. Originally, the source was polonium 210, but the half-life of polonium 210 is only 138 days, and so its intensity drops off each year to ⅛ of what it was the year before. Lead 210 has a half-life of 25 years and decays to polonium 210, which is the element that produces the visible tracks in the cloud chamber. The lead 210 replenishes the polonium 210 almost as fast as it disintegrates. Don't confuse your students with this detailed information. The important thing is that the tracks are produced in the cloud chamber by the decay of polonium. The radioactive source should be weak enough so that individual events can be clearly observed and their randomness noted.

The dark felt band encircling the inside of the top of the chamber must be evenly soaked with about one eyedropperful of methanol or isopropyl alcohol.

The cloud chamber is placed on a block of solid carbon dioxide (Dry Ice), which usually is available from large dairy or ice-cream stores. Since the sublimation temperature is −78°C, the Dry Ice should be handled with gloves or a cloth. It can easily be broken up with a hammer or cut with a saw into a flat piece like the one shown in Fig. 7.8.

If Dry Ice is not commercially available, a carbon dioxide fire extinguisher can supply solid carbon dioxide for the experiment. The fire extinguisher must be obtained specifically for this purpose; a unit assigned permanently to the classroom for protection must not be used.

The following procedure has proved very satisfactory. Cut out a square block about 1½ inches thick from a slab of Styrofoam or a dry sponge. The square should be at least as large as the horn opening of the extinguisher and not smaller than the block of Dry Ice shown in Fig. 7.8. Set the horn of the extinguisher squarely on top of the Styrofoam block and, holding it down firmly, release carbon dioxide for about 4 seconds. The tank of the extinguisher should not be tilted more than about 60°. If it is tilted more than this and the tank is partly empty, only gaseous CO_2 will be expelled and no more Dry Ice will form. Remove the extinguisher, smooth out the solidified carbon dioxide on the block with a ruler edge, and place the cloud chamber on the layer of Dry Ice.

It will take 5 to 10 minutes for the cloud chamber to become cold enough to supersaturate the alcohol vapor near the bottom and allow fog tracks to form. Illuminating the chamber from the side with a beam of light from a slide projector or a flashlight will help considerably in making the fog tracks visible against the black bottom. A layer of solidified carbon dioxide about ¼ inch thick on the Styrofoam will keep the cloud chamber going for about 20 minutes.

We repeat what we stated in the overview of the chapter: The second purpose of studying radioactivity is to suggest the particle nature of matter. This section contains the essential points for this purpose; these are the points the students should have in mind when they begin the study of the atomic model of matter introduced in the next chapter. Resist the temptation to expand the treatment of the subject beyond what is given here.

CHAPTER 7—ANSWERS TO PROBLEMS

Sec.	Easy	Medium	Hard	Class Discussion	Home or Lab
1	1, 2				
2		3, 4, 5, 20	18	5	20
3	6†				
4	8†	7			
5, 6		9, 10			19
7	11†, 12†, 21, 22	23			
8	15	13†, 14			
9	16	17, 24			

1 **What would happen if you heated the solid that you recovered in this experiment?**

The solid has all the properties of copper; it is copper. Therefore, it would once again turn black and gain mass by combining with oxygen from the air.

2 **How do you expect the total mass of solid recovered in Expts. 6.7 and 7.1 to compare with the initial mass of copper you heated in Expt. 6.6?**

If the solids recovered in Expts. 6.7 and 7.1 are pure copper, you expect that the sum of their masses should be the same as the initial mass of copper in Expt. 6.6.

3 **While on a class field trip a student found a shiny rock that appeared to be a metal. When she returned to the classroom, she heated the rock and found that it lost mass. Could this rock be an element? Explain your answer.**

The rock could not be an element. Because the rock lost mass when it was heated, it could have been a compound that decomposed into two simpler substances—into a solid and a gas.
  Either or both the remaining solid and the gas might be elements.
  There is another possibility. The rock could be a mixture of a solid (element or compound) and absorbed water or other liquid. When heated, the liquid could have vaporized and been driven off, causing the original solid to lose mass.
  In any case, the original solid was not a *pure* element.

4 **How do you know that water, zinc chloride, and sodium chlorate are not elements?**

In Expt. 6.1 sodium chlorate was decomposed by heat into simpler substances; in Expt. 6.2 water was decomposed by electrolysis into simpler substances. Neither can be an element. In Expt. 6.4, the mass of zinc chloride produced is greater than the mass of the zinc from which it was made; something was added to the zinc in the process. From this we infer that zinc chloride can be separated into simpler substances and, therefore, cannot be an element.

5   Hydrogen chloride—a gaseous pure substance—can be decomposed into two different gases, each of which acts like a pure substance. On the basis of this evidence alone:
a)   Can hydrogen chloride be an element?
b)   Can either of the other two gases be an element?
c)   Can you be sure that any of the pure substances mentioned is an element?

a)   The hydrogen chloride gas cannot be an element because it has been separated into two different substances.
b)   Yes.
c)   We cannot be sure that either of the gases produced is an element without further evidence, but we are certain that hydrogen chloride is not.

6† Lavoisier's list of the elements (Table 7.1) included *magnésie*, *baryte*, *alumine*, and *silice*, all of which are now known to be compounds. What do you think are the elements in each of these compounds?

The modern English names listed in the table suggest that the compounds contain the following elements:

Magnésie: magnesium, oxygen   *Alumine*: aluminium, oxygen
Baryte: barium, oxygen   *Silice*:   silicon, oxygen

7   When you hold a small amount of sodium chloride in a flame, you observe that the flame is strongly colored yellow. What could you do to be sure that the color is due to the sodium and not to chlorine?

Examine the color of the flame obtained when compounds of sodium other than sodium chloride are tested. The yellow color is common to all.

8† If you spill a few drops of soup or milk on a pale-blue gas flame when cooking, the flame changes to a mixture of colors with yellow being the most intense color. How do you explain this fact?

Minerals in the soup or milk contain metals that give characteristic colors to the flame. Since sodium compounds are almost universally present in foods, the yellow color of the sodium flame is almost always observed in cases like this.

9 Figure A shows some spectra observed with the same spectroscope that was used to obtain the spectra illustrated in Fig. 7.2. What elements can you identify in (*a*)? In (*b*)?

A direct comparison of these spectra is possible because they are to the same scale.
a) This spectrum is due to a mixture of calcium and lithium.
b) This spectrum is due to a mixture of strontium and some element or elements not identified.

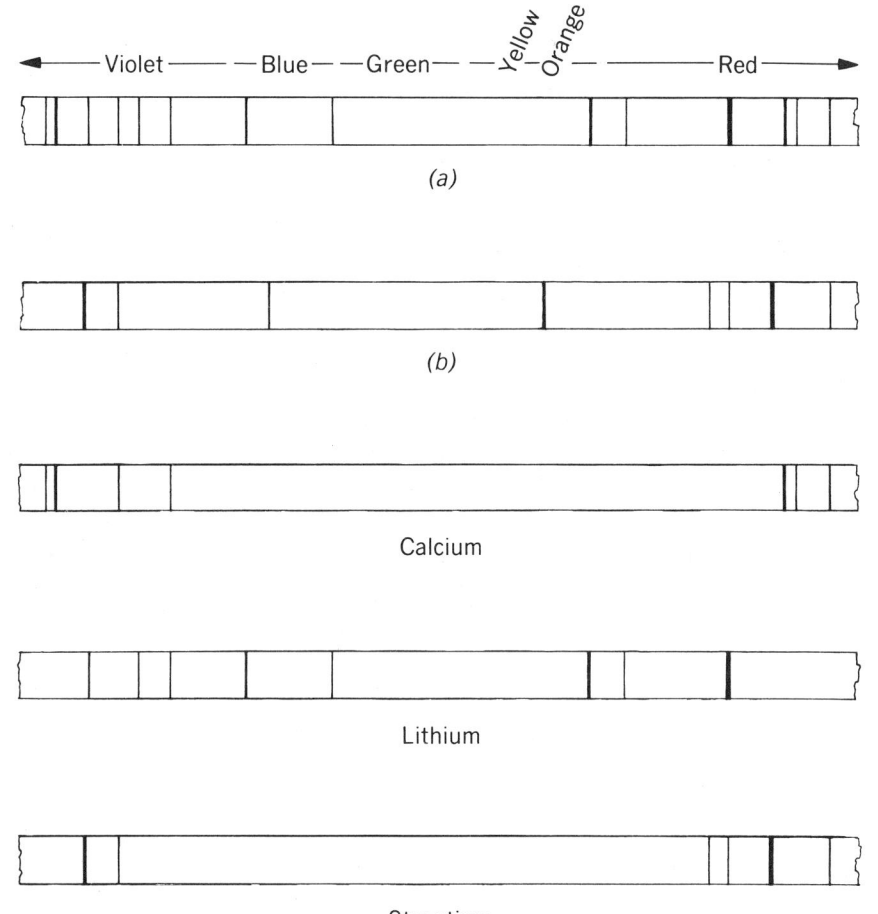

◄──── Violet ──── Blue ── Green ── Yellow/Orange ──── Red ────►

(a)

(b)

Calcium

Lithium

Strontium

Figure A

10 Suppose that you were given a sample of a substance. How would you try to find out if the substance were an element, a compound, or a mixture?

This problem may well serve as a starting point for a review of the course up to this point. A close visual inspection of the substance may help to select the experiments to do first. Also, in the case of a solid, it might be wise to grind it into a powder if possible. You can use the following methods to try to answer the question.

Solids	Liquids
Fractional crystallization	Chromatography
Melting point	Fractional distillation
Reaction with known substances	Decomposition by heating
	Reaction with known substances
Spectral analysis	Spectral analysis

There are only four elements that are liquids at or near room temperature—mercury (m.p. −38.9°C), bromine (m.p. −7.2°C), cesium (m.p. 28.5°C), and gallium (m.p. 29.8°C).

Gases

Boiling point
Liquefaction and fractional distillation
Reaction with known substances
Spectral analysis in an electrical discharge tube

It is often difficult to say if the original substance was a mixture, a compound, or an element. In general, mixtures separate, and their ingredients can be recombined more easily than compounds. The ingredients of a mixture can be recombined in almost any proportion, whereas elements combine to form compounds in a constant proportion for a given compound. Caution must be observed here when considering solutions that saturate at a fixed ratio of solid to liquid and compounds that do not appear to show constant composition.

If none of the separation processes listed above produces a separation, and if spectral analysis shows the presence of only one element, the substance may be an element.

11† Elements *X*, *Y*, and *Z* form compounds *XY*, *XZ*, and *YZ*. Compounds *XY* and *YZ* are radioactive, but *XZ* is not. Which element is radioactive?

Y

12† A piece of magnesium placed in hydrochloric acid causes hydrogen to be released. Evaporation of the resulting solution leaves behind a white solid. A similar reaction occurs when a piece of uranium is placed in hydrochloric acid. Would you expect this white solid to be radioactive?

Yes.

13† A single sample of uranium nitrate was left on a piece of photographic film for a week. During the week the sample was moved twice to new spots on the film. When the film was developed, the three "exposed" areas were found to vary in intensity. What can you conclude about (a) the temperature variation during the week and (b) the length of time the sample was in each position?

a) Nothing.
b) The sample remained the longest on the most intensely exposed area.

14 What are the two most important differences between the following two reactions?
a) Water → hydrogen + oxygen
b) Polonium → lead + helium

The most significant difference between the two reactions is that we can control the rate at which (*a*) takes place, but we cannot control (*b*). If the number of batteries used for the electrolysis is increased, the water decomposes more rapidly. Or, if the amount of acid in the water is decreased, the water decomposes more slowly.

Another significant difference between the two reactions is the easy recombination of hydrogen and oxygen to form water and the practical impossibility of recombining lead and helium to make polonium.

15 A radioactive sample at 20°C is placed near a device that counts the radiation coming from the sample. The counter records 1.0×10^2 counts/min. The temperature of the sample is then raised to 100°C. What does the counter then record?

Since the intensity of the radiation is independent of temperature, the counter will normally record an average of 1.0×10^2 counts/min.

16 You look through a microscope at a photographic plate that has been exposed for 30 hours to a needle tip containing a tiny bit of radioactive substance. You count 563 dots. How many dots would you expect to count if the film were exposed for only 10 hours?

If we assume that the rate of decomposition is practically constant during the 30 hours, then in one-third the time we should expect only about one third as many dots, or close to 188. Owing to the randomness of the radioactive disintegrations, we might get a few less or a few more than 188 dots.

17 The number of black dots on the negatives in Fig. 7.7 is related to the exposure time. How do you suppose this relationship would work after very long exposure times?

The proportionality of the number of dots to the exposure time cannot hold forever; after all the radioactive element has decayed, waiting longer will not produce new dots. If measured over a long enough period, the production of dots per hour will decrease with time. The rate of this decrease depends on the radioactive element used.

18 When magnesium is put into hydrochloric acid, the metal reacts, a gas bubbles off, and a white solid is left behind after evaporation.
a) On the basis of this information alone, can you be sure which of the pure substances mentioned are compounds and which are elements?
b) We then find that the gas reacts just like hydrogen gas; it has the same characteristic properties. Also we find that hydrochloric acid can be decomposed into hydrogen and chlorine. Can the remaining solid be an element?
c) From the table of known elements (Table 7.2, page 110), which of the substances mentioned above are elements and which are compounds?

a) On the basis of the information given, we have no way of deciding which, if any, of the substances are elements.
b) The remaining solid is not an element, because all the elements involved are now known, as a result of the information given in (*b*), and only magnesium is not a white solid.
c) Elements: magnesium, hydrogen.
    Compounds: hydrochloric acid, magnesium chloride.

19 Use your grating spectroscope to examine the spectra of street lights, "neon" store lights, and any other light sources you can find (except the sun, which is too bright and will damage your eye if you look directly at it). Record your observations and compare them with those of your classmates.

This activity is not very effective unless single light sources can be examined in a darkened room or unless store signs can be examined at night.
    Neon bulbs are fine; the spectrum is rich in lines, and the appearance of green lines is unexpected. Many store signs are called "neon" when in fact they are fluorescent lights or contain gases other than neon.
    With fluorescent lights, the strong green and blue lines of mercury can be clearly observed against the complete visible spectrum in the background.

20 a) If you heat a piece of blue vitriol (a blue solid), it loses mass and changes to a white powder. Which of the two substances might be an element?

b) If you dissolve the white powder in water and place an iron nail (iron is an element) in the solution, the nail will become coated with a thin layer of copper. What do you now conclude about the two substances in part (a)? Is either of them an element?

a) The white powder might be an element. The blue vitriol could not be, since it was separated into simpler substances—one a white powder, the other possibly a gas (actually, it is water vapor).
b) Both the blue vitriol and the white powder contained copper. Since copper is an element, and neither of the two solids has the properties of copper, the two solids must be some combination of copper with at least one other substance. Neither of the two substances can be an element.

21 A plant absorbs various substances through its roots. Different elements of these substances concentrate in different parts of the plant. Suppose one of these elements is radioactive. How would you determine in what parts of the plant it concentrates?

Various parts of the plant could be left for a suitable time in contact with photographic film. The exposure of the film would reveal the location of the parts where the radioactive element is concentrated. The various parts of the plant could also be examined with a Geiger counter. The radioactive element is concentrated in those parts that cause the Geiger counter to click.

22 In the film loop "Radioactive Substances I" the Geiger counter counted much faster when box A was placed next to it than when box E was. Is this what you would have expected on the basis of the brightness of the white patches in Fig. 7.4? Why or why not?

Yes. The effect of box A on the film was greater than the effect of box E. Since the two boxes were in contact with the film the same length of time, the radioactivity of the substance in box A must be greater than that of the substance in box E. Therefore, you would expect the counter to count faster when box A is near it.

23 You saw in the film loop "Radioactive Substances I" that a Geiger counter counts something even if it is not placed near a radioactive source. Taking this fact into account, how would you measure the number of clicks per hour given off by a radioactive source?

You would record the number of counts per minute made by the Geiger counter when no radioactive source was near it. This is called the background count. The background count per minute multiplied by 60 gives the background count per hour. This you would subtract from the total count in an hour to get the count per hour caused by the radioactive source alone.

24 The film loop "Radioactive Substances II" shows tracks being produced in a cloud chamber. From what you see in the loop, can you be sure that the tracks start at the source rather than end there? What does this tell you about the speed of the particles that leave the tracks?

As far as the eye can see, the whole track appears all at once, and you cannot tell which way any one particle is going. The particles must be moving at high speed. (Of course, the fact that all tracks radiate from the source strongly suggests that the particles start from the source.)

The Atomic Model of Matter

8

Overview of the Chapter

We now begin the consideration of the atomic model of matter, which will be continued in the next two chapters. After the general principles of a model are introduced, the atomic model is suggested.

To emphasize the process of building the atomic model and its underlying logical structure, a summary of this chapter is provided in Table A. In the left-hand column of this table are observations; in the right-hand column is the evolving description of the model. Arrows show where the observations were used to build the model and where the model correlated several observations or suggested new ones. The items on the left do not prove those on the right—they merely suggest them. The value of the model is judged by its usefulness.

An experiment with rings and fasteners demonstrates that the law of constant proportions is consistent with a particle model. A second experiment with rings and fasteners allows us to predict the law of multiple proportions, and an experiment with two copper chlorides supports this prediction.

A final section discusses the existence of molecules in gases and solids and relates it to observed differences in the compressibilities of solids, liquids, and gases.

If the end of the school year is approaching and you anticipate that you will not be able to complete the course, you may wish to eliminate parts of Chap. 8. Many teachers have eliminated the study of multiple proportions (Sec. 8.5 through Sec. 8.8) in favor of considering a larger portion of the remainder of the course.

The suggested schedule for this chapter is:

Sections 1-2 (one experiment, probs. 1-3, 20-21)	1 period
Sections 3-5 (two experiments, probs. 4-10, 22)	4 periods
Sections 6-8 (one experiment, probs. 11-17, 23, 26)	5 periods
Sections 9-10 (probs. 18-19, 24-25, 27-28)	1 period
Film: *Definite and Multiple Proportions*	2 periods
Achievement Test No. 4	2 periods
Total	15 periods

Table A The Atomic Model of Matter

Observations	Model
I Direct counting with a Geiger counter, discrete dots on film, and individual tracks in cloud chamber.	Elements consist of discrete particles or atoms.
II Atoms cannot be seen with a high-powered microscope.	Atoms must be very small and therefore there must be very many of them.
III Elements have characteristic properties.	All atoms of the same element are identical. Atoms of different elements are different from one another.
IV Elements form compounds that have their own characteristic properites, yet it is possible to extract the elements again from their compounds. Mass is conserved in chemical reactions. Spectral analysis. Law of definite proportions. Law of multiple proportions.	Atoms combine to form special patterns that are different from mere mixtures of atoms. The pattern and ratio of atoms in a given compound are always the same. In forming these patterns, or in breaking them up, the number of atoms and their masses do not change. There are no fractions of atoms in compounds.

The Atomic Model of Matter   157

Observations	Model
V Elements and compounds are substantially less dense as gases than as solids or liquids. Gases are easily compressible; solids and liquids are not.	Atoms behave like small, hard objects. In a gas they are far apart. In a solid they "touch" one another. If the atoms in a gas are arranged in clusters (or molecules) that contain a fixed number of atoms, the distance between the centers of atoms in the molecules of a gas is much less than the distance between the molecules. Molecules of some liquids and solids have a structure such that when they are "touching," as they are in liquids and solids, there is a relatively large volume of empty space between them (like a jumbled box of matches.)
Liquids and solids that have a molecular structure are more compressible than those without such a structure.	
VI One element changes into two other elements in radioactive decay.	A radioactive element disintegrates into atoms of other elements.

A MODEL   8.1

Treat this section as an introduction to the experiment in the next section, which will further amplify the idea of a model.

EXPERIMENT: A BLACK BOX   8.2

Students require guidance in the profitable ways to observe the properties of the boxes, in the proper statement of their observations, and in constructing the set of ideas about the internal construction of the boxes that constitute the model.

In the pre-lab impress upon your students the need to make careful observations and to record them. Roughly shaking the boxes will not be nearly as revealing as carefully tilting them.

Each group of four students should be given several identical boxes, and the observations, model building, and predictions should be done by the group.

The "black boxes" contain metal or plastic rods, passing through the sides of a cardboard box, and arrangements of metal washers either riding on the rods or loose in the box. One such arrangement is illustrated in Fig. I. The "black boxes" are available from suppliers, or can be made from knitting needles and small gift boxes, obtainable at gift or variety stores.

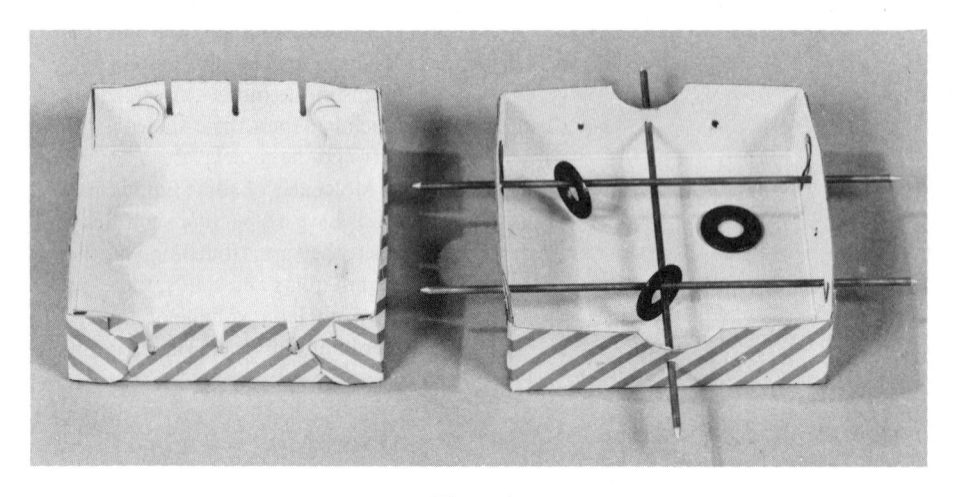

Figure I

Many configurations are possible, and a few successful ones are shown in Fig. II.

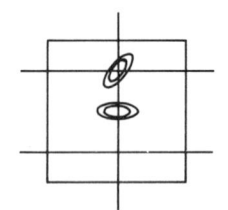

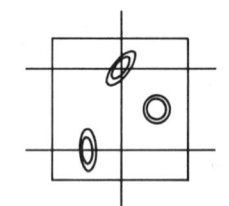

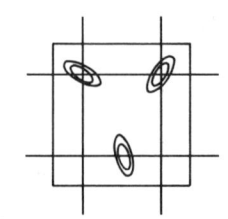

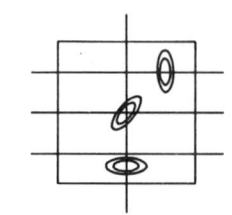

Figure II

Mark each box with a letter, and number the rod positions so that your students can easily identify them.

A sample set of observations might be as follows:

Observations—Box A

1. Tilted the box in several directions; a sliding noise is heard coming from the bottom of the box, followed by a thump on the lower side.

2. Tilted the box from the horizontal around an axis parallel to that of the two parallel rods; a sliding noise comes from the bottom surface of

the box as before. A very slight click occurs almost immediately, even when the tilt is too small to cause the sliding noise.

3. Tilted the box from the horizontal around an axis parallel to the single rod; a sliding noise from the bottom of the box as before, a prompt, very slight click as before; a sliding noise apparently coming from one of the rods, followed by a thump on the side of the box. Tilting in the opposite direction produces the same sliding noises and slight click, except that one of the sliding noises ends with a click instead of a thump.

4. Some kinds of rapid twisting produce a rolling sound, somewhat as a coin does when it is dropped on the floor and rolls around briefly on its rim before coming to rest.

After each student records his observations, he must decide on a model for the box. A student might start off by saying, "The box contains a loose, disk-shaped solid object." This idea would be consistent with observations 1 and 4. The student would continue until the model accounted for all of his or her observations. The model should be expressed in the form of a diagram showing the arrangement of rods and objects. When your students have formulated their models, they can make predictions about what will happen when a particular rod is removed and what new properties the boxes will then have. Unexpected results will require modifications of the model and more testing.

A group may decide to pull out rod No. 1. If their model is correct, they predict that a thump should be heard when the rod is withdrawn. This prediction is tested and found to agree with the model. A second prediction might indicate that the withdrawal of rods 2 and 3 should have no effect. When this experiment is tried, a thump is again heard. This observation should suggest a new model to the group—perhaps a disk-like object with a hole in it is to be found at the intersection of rods 2 and 3.

Do not allow the box to be opened at any time, even after the experiment is completed. The contents thus cannot be positively identified, but can be described only in general terms. This corresponds to physical models where the exact "contents" of a system can never be completely checked.

Encourage your students to construct models of their own at home and bring them to school for their classmates to investigate. Impress upon them that you are not after a "What's-in-the-box?" game, but rather a "box" that will suggest a series of logically connected investigations leading to a testable prediction. The most interesting and valuable "boxes" will be those that lead to plausible predictions, which will, nevertheless, be affected by further testing.

THE ATOMIC MODEL OF MATTER 8.3

As in the case of the "black box" model, the atomic model must first account for the properties of matter that we have already encountered. To do this for the properties discussed in this section, a set of assumptions is introduced into the model to account for the specific properties. These are outlined in parts III and IV of Table A. Note that the arrows run both ways in this table. The properties discovered in the laboratory suggest certain assumptions about the model, which in turn may explain other properties of matter already observed.

Some of these assumptions may suggest or predict a property of matter that has not yet been observed. This aspect of the atomic model will be discussed in Sec. 8.6.

Outline the appropriate sections of Table A on the chalkboard with specific examples, as follows:

Observation

1. Elements have characteristic properties.

Carbon is a pure substance with certain characteristic properties (density of 1.9 g/cm^3, etc.); oxygen is a pure substance with certain characteristic properties (density of 1.4×10^{-3} g/cm^3, supports combustion, etc.).

2. Elements form compounds that have their own characteristic properties; yet it is possible to extract the elements again from their compounds.

Carbon and oxygen unite to form a compound, carbon dioxide, that has quite different characteristic properties (density of 2.0×10^{-3} g/cm^3, does not support combustion or burn). Carbon dioxide can be decomposed into carbon and oxygen.

The Model

1. All atoms of the same element are identical. Atoms of different elements are different from one another.

All the carbon atoms are alike and all the oxygen atoms are alike, but the oxygen atoms are different from the carbon atoms.

2. Atoms combine to form special patterns that are different from mere mixtures of atoms. The pattern and ratio of atoms in a given compound are always the same.

A given number (unknown at this point) of carbon atoms are attached to a given number of oxygen atoms to form a particular arrangement of carbon and oxygen atoms in carbon dioxide. Atoms of elements do not disappear in a compound and can be recovered by decomposing the compound.

Be careful not to carry the examples too far. There is no basis for discussing in this section the size, shape, number, mass, internal structure, or other properties of individual atoms, or even the existence of molecules.

8.4 "EXPERIMENT": FASTENERS AND RINGS; CONSTANT COMPOSITION

The use of objects such as brass paper fasteners and rubber grommets in this experiment is intended to help students understand the abstract atomic model of matter. By combining real objects to make "compounds," they will see more concretely the relationship between the model and the law of constant proportions.

You may find it wise, in the pre-lab, to remind some students, as is done in the text, that atoms are not miniature fasteners and rings—that these objects are used only because they serve well to illustrate the combination of atoms into compounds.

The operations performed in this experiment are analogous to those used in the synthesis of zinc chloride, Expt. 6.4. There a measured mass of zinc was reacted with an acid to form zinc chloride. Any unreacted zinc was massed, and the mass of zinc that reacted was compared with the mass of zinc chloride formed. In this experiment a measured mass of Fs is "reacted" with R to form FsR. Any unreacted Fs is massed, and the mass of Fs that reacted is compared with the mass of FsR formed.

The Experiment

Give half of your students between 10 and 50 brass fasteners and enough rubber rings so that some will have an excess of R and others an excess of Fs in synthesizing FsR. To demonstrate the law of constant proportions, give different groups different total amounts, so that a comparison of class results will show that the mass ratios found are independent of the mass of starting materials and the substance present in excess.

If possible, give some of your students between 10 and 50 brass fasteners and enough *metal nuts*, whose individual mass is different from that of a rubber ring (and a fastener), to enable them to proceed as described above. If you do this, two constant ratios will be found, one ratio by half of the class and the other by the other half, showing that each compound will have a different constant proportion.

In making FsR (or FsNt) to find the mass ratio of Fs to FsR (or FsNt) in the "compound," make sure your students make as much of the compound as their supply allows, so that the "reaction" is as much like a real reaction as possible. They may wish to mass only one ring and one fastener and figure out the results from that alone. However, this is an unrealistic illustration of a real reaction for the following reasons: (1) We cannot put a single atom on a balance and mass it, and (2) it is impossible to react only one real atom with one other atom. Real reactions, even on a small scale, involve millions of atoms.

In this experiment try to get your students into a frame of mind that allows them to follow all aspects of the parallelism of their experiment with chemical experiments involving real reactions.

Sample Data (using 25 fasteners and 25 rings)

Mass of Fs	26.22 g
Mass of FsR	26.83 g
Mass of excess Fs	4.37 g
Mass of reacted Fs (26.22 − 4.37)	21.85 g
Mass of reacted R (26.83 − 21.85)	4.98 g

Ratio of reacted R to product:

$$\frac{4.98 \text{ g}}{26.83 \text{ g}} = 0.186$$

Class data for the two ratios, R/FsR and Nt/FsNt, can be tabulated, and the results should be summarized in a histogram.

162 The Atomic Model of Matter

Answers to Questions

As shown by class results, the ratio of the mass of R to product does not depend on the size of the sample. The fastener-and-ring illustration of the model agrees with the law of constant proportions. If heavier rings are used (or if nuts are used in place of rings), the ratio is not the same, but it is constant regardless of quantity for any given combination.

Decomposition of the compound into "elements" would show that the model agrees with the law of conservation of mass.

Apparatus and Materials

Balance
10-50 Brass paper fasteners
15-30 Rubber rings (grommets), *or*
15-30 Metal nuts

8.5 "EXPERIMENT": SOME OTHER COMPOUNDS OF FS AND R

Two compounds of Fs and R are decomposed, and the masses of R that would have combined with 100 g of Fs are determined for each compound. These masses are compared and found to be in a ratio of small whole numbers.

The Experiment

Give each group enough fasteners and rings to make between 15 and 25 "clusters" of both FsR_2 and FsR_3.

Sample Data

Compound FsR_2 (20 fasteners and 40 rings)

| Mass of R | 7.96 g |
| Mass of Fs | 17.46 g |

Mass of R combined with 100 g of Fs:

$$\frac{7.96 \text{ g}}{17.46 \text{ g}} \times 100 \text{ g} = 45.6 \text{ g}$$

Compound FsR (25 fasteners and 25 rings)

Using the data from the previous experiment:
| Mass of R | 4.98 g |
| Mass of Fs | 21.85 g |

Mass of R combined with 100 g of Fs:

$$\frac{4.98 \text{ g}}{21.85 \text{ g}} \times 100 \text{ g} = 22.8 \text{ g}$$

Compound FsR_3 (20 fasteners and 60 rings)

| Mass of R | 11.93 g |
| Mass of Fs | 17.49 g |

Mass of R combined with 100 g of Fs:

$$\frac{11.93 \text{ g}}{17.49 \text{ g}} \times 100 \text{ g} = 68.2 \text{ g}$$

Summary of Results

Compound	Mass of R Combined with 100 g of Fs	Mass Ratios
FsR	$M_1 = 22.8$ g	
FsR$_2$	$M_2 = 45.6$ g	$M_2/M_1 = 2/1$
FsR$_3$	$M_3 = 68.2$ g	$M_3/M_1 = 3/1$

The data, using different amounts of the compounds, will show that these ratios do not depend on the mass of the compounds made.

Apparatus and Materials

Balance
15-25 Brass fasteners
30-75 Rubber rings (grommets)

A PREDICTION FROM THE ATOMIC MODEL OF MATTER 8.6

This section is a reading assignment. Discuss it thoroughly in the pre-lab session for the next experiment, since it must be understood if the results of the next experiment are to be interpreted correctly and the law of multiple proportions understood. The reasoning parallels that of the last experiment on fasteners and rings.

EXPERIMENT: TWO COMPOUNDS OF COPPER 8.7

The reaction in Expt. 6.4 showed a constant mass ratio of starting material used to product formed, no matter what amounts of substances were initially put together. This law of constant proportions is essential to the idea that a compound is not simply a mixture of various elements in any proportion. Students learned that elements combine to form a compound only in a specific ratio.

The model for matter predicts, in Sec. 8.6, that for two compounds made only of the elements copper and chlorine, we might expect the ratio of the mass of chlorine combining with 100 g of copper in one compound to the mass of chlorine that is combined with 100 g in the other compound to be 2, 3, 4, or some other ratio of small integers. The experiment is designed to test this prediction and is worth the time it takes to do it.

Be sure to point out to your students that this is a difficult experiment, and urge them to work carefully. The experiment involves many steps in which the experiment can be ruined. In order to be reasonably sure of success, it is best to give your students plenty of time.

The best class results can be obtained when half the class uses the brown powder and the other half the green powder on the first run. This will result in more equal spreads of the masses of copper from the two chlorides. Allow three or four class periods, depending on the method of drying used. A fair amount of pre-lab discussion is necessary, but just as important is the time required for you to check your students continually as they work. When they hand in their final results, check them immediately step by step. Errors are often found in the calculations, and you will wish to have them corrected before the results of the class are analyzed with a histogram in the post-lab discussion.

The first problem is the drying of the original sample. It is necessary that the chlorides be free of all absorbed moisture as well as all water of hydration. Copper (II) chloride (cupric chloride) is hygroscopic and readily forms a blue-green hydrate. When thoroughly dry, it has a brownish-yellow color. Copper (I) chloride (cuprous chloride) is not very hygroscopic, but will absorb some moisture if left uncovered.* Dry the material in advance. Both chlorides can be dried at the same time, using one of the following methods:

1. Using a large mortar and pestle, thoroughly pulverize about 60 g of material (for a class of 12 groups). Spread the powdered material thinly on notebook paper. The substance can now be dried under an infrared heat lamp placed about 8 inches above the paper. The temperature should not rise above about 110°C. The blue-green hydrated chloride (copper II) chloride) will rapidly change to a brownish-yellow color. Stir the material every ten minutes or so. The drying will take about an hour. The dried chlorides should be stored in tightly stoppered bottles until used.

2. If a drying oven is available, the powder may be crushed, spread out, and dried at a temperature of 100 ± 10°C for an hour or so in the oven.

3. Lacking a drying oven or an infrared heat lamp, you can make a substitute oven (Fig. III). A corrugated cardboard carton is placed on one side so that the inside is made accessible by lifting one of the cover flaps. Four 60-watt light bulbs in surface mounting sockets (the sockets do not have to be fastened to the bottom) are wired in parallel to form a rectangle and placed on the bottom of the box. The wire screen rests on or is just over the bulbs. Within an hour after the lights are turned on, the temperature will reach about 90°C in the closed box.

This box, if needed, may also be used later for drying your students' copper samples. (The screen shelf is large enough to accommodate at least 18 watch glasses.)

For an 18 in. × 12 in. box the screen should be 18½ in. × 12½ in. All four edges of the screen should be folded upward so that it just fits snugly in the box. Before folding, cut a quarter-inch square out of each corner of the screen. Next cut two boards about 1 inch shorter than the longest dimension of the box. To fold the screen, clamp the two boards

---

*Pure copper (I) chloride is white. It is very difficult to prepare and keep pure, since contact with even a little moisture will convert some of it to copper (II) chloride. The material your students use is light green because of the presence of a very small amount of hydrated copper (II) chloride even when thoroughly dried. However, if reasonable precautions are taken to keep the copper (I) chloride dry, there will not be enough of the green chloride present to affect the results.

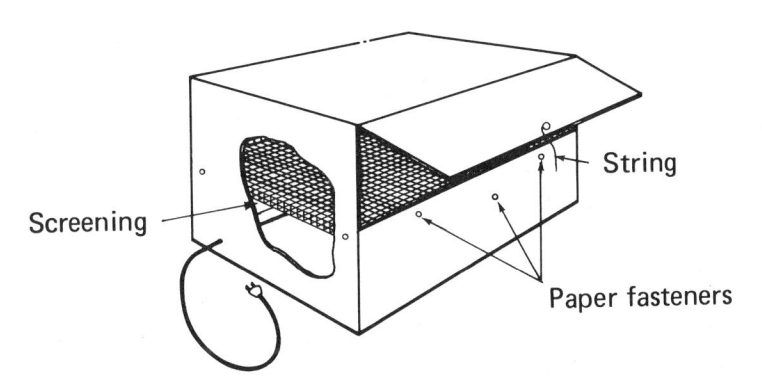

Figure III

together with C-clamps with a quarter-inch of the screen edge between them. The screen can now be bent by pushing down on a table. If necessary, you can finish by hammering the screen flat along the bend. The screening, with folded edges up, is secured to the walls of the box with eight paper fasteners as shown in the figure.

The inner cover flaps are mostly cut away as shown in Fig. III. The bottom outer cover flap is then secured with paper fasteners to the remains of the inner cover flaps as shown. A pair of paper fasteners and a piece of string serve to keep the upper cover flap closed.

Parts List: Drying Oven

Cardboard carton
 18 in. × 12 in. × 12 in.
 (or larger)
One-quarter-inch mesh metal
 screening 18½ in. × 12½ in.
 (or larger) to fit the box
 when folded

12 Paper fasteners (like those
 used in Expt. 8.4)
String
4 Surface-mounting light-bulb
 sockets
4 60-watt bulbs
Appliance wire (about 10 ft)
Appliance plug

The Experiment

Have different groups mass out different amounts of the brown powder (copper (II) chloride) in the 2-g to 4-g range so that class results, when compared, will show that the ratio does not depend on the mass used. The powder is poured directly onto a massed watch glass or into a paper cup until it roughly balances the required mass. Then the balance can be adjusted to determine accurately the mass of powder. The same massed watch glass can be used later to dry the copper over a water bath. The brownish-yellow copper (II) chloride will readily dissolve, but the other chloride is only slightly soluble in water. However, it will begin to dissolve as soon as the aluminum starts to react with the hot solution and replaces the copper in the copper chloride that is in solution.

If the coiled aluminum strips (Fig. IV) are placed in the solutions before heating, the first copper that forms will stick so hard to the aluminum that it may be impossible to get it off. Therefore, the aluminum should not be added until the solutions are heated to around 60°C. (If you wish, you may issue thermometers to your students so that they can be sure that the water is at the right temperature before introducing the aluminum.

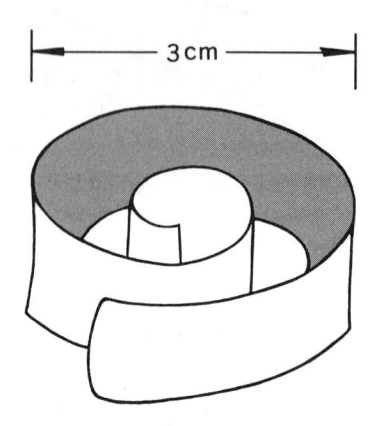

Figure IV

The rate of the reaction depends on the surface area of the aluminum that is in contact with the solution. Therefore, in the case of the copper (II) chloride, the reaction will go to completion most quickly if all the aluminum is submerged and if the solution is stirred constantly. Gas bubbles will be given off, while at the same time a reddish-brown precipitate of pure copper forms on the surface of the coil. The spongy copper precipitate will usually coagulate and separate from the aluminum in about 5 to 10 minutes. If it does not, the copper precipitate can be dislodged from the aluminum by agitating the coil with a glass stirring rod or scoopula. When the color disappears from the solution, the reaction is essentially over.

The surface of the aluminum strip is protected thoroughly by an oxide coating, which re-forms very rapidly in air if scraped off. When the reaction with copper chloride begins, an acid solution forms, the oxide coating is eaten away, and the freshly exposed surface of pure aluminum can then itself react with the acid solution to release hydrogen from the acid. This secondary reaction will not interfere with the precipitation of copper, but hydrogen may continue to be given off after the reaction with copper is complete. Have your students go ahead with the rest of the experiment, even though the solution may still be fizzing.

It is very important that, when the reaction that precipitates copper is complete, the copper be thoroughly washed twice with water and once with alcohol and that none of it be lost in decanting the wash water or in transferring the copper to the pre-massed watch glass.

After the copper is washed, it can be transferred to the massed watch glass or massing cup with a scoopula.

Possible time schedules for this experiment are listed below. *Do not hurry your students!* It is important to allow enough time to ensure accurate results.

First period. Pre-lab discussion and completion of the experiment with the first chloride to the point where the alcohol-washed copper is on the watch glass.

Second period. Massing the dried copper. Completion of the experiment with the second chloride to the point where the alcohol-washed copper is on the watch glass.

Third period. Massing the dried copper. Calculations, post-lab.

Sample Data

	Masses (g)	
	Brown chloride	Green chloride
Copper chloride used	2.00	2.00
Copper produced	0.96	1.28
Chlorine in copper chloride	1.04	0.72
Chlorine combined with 100 g of copper	108	56

Be sure to plot the data for the entire class on a histogram. Figure V shows a sample histogram plotted from the combined data of three classes. From it one may conclude that approximately 110 g of chlorine combine with 100 g of copper in the brown chloride, while approximately 55 g of chlorine combine with 100 g of copper in the green chloride, giving a ratio of 110/55 = 2.0.

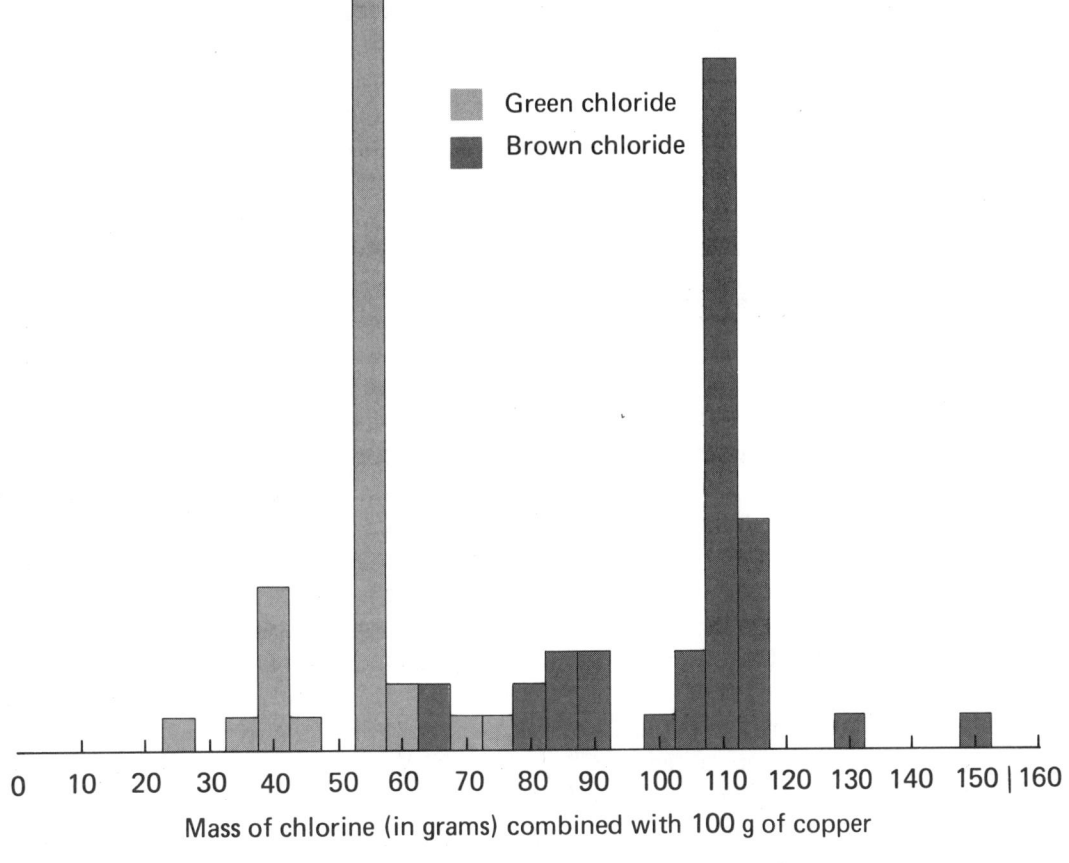

Figure V

168 The Atomic Model of Matter

Answers to Questions

The sample of copper is dry enough if the mass does not change by more than about 0.01 g after 10 additional minutes of heating.

The texture of the copper is different from that of common samples of copper, but squeezing and hammering make it appear more like common, solid copper.

See the Sample Data and the histogram for typical masses of copper and chlorine and for the analysis of the experiment.

Twice as much chlorine combines with a given mass of copper in the brown compound as combines with the same mass of copper in the green compound. There are twice as many chlorine atoms for every copper atom in the brown compound as there are in the green compound.

(Note that there is no way of telling from this experiment how many chlorine atoms there are for each copper atom in either of the compounds.)

Apparatus and Materials

Balance
Burner stand
Alcohol burner
Beaker (100 cm^3)
Beaker (250 cm^3)
Graduated cylinder (50 cm^3)
Watch glass
Glass stirring rod
Scoopula
Coiled aluminum strip
 (1 cm × 10 cm × 0.05 cm, about 1.4 g)

Copper (II) chloride
 (cupric chloride) 2-4 g*
Copper (I) chloride
 (cuprous chloride) 2-4 g*
Burner fuel
Matches
Water
Drying oven or equivalent
 (optional)
Thermometer (optional)

8.8 THE LAW OF MULTIPLE PROPORTIONS

There are two main ideas in this section which warrant thorough class discussion. The first relates to Table 8.1 as contrasted with Tables 8.2 and 8.3. Table 8.1 presents conclusions from the atomic model. We actually take fasteners and rings, put them together in specified combinations, and measure the mass ratios. Tables 8.2 and 8.3, on the other hand, present the results of measurements made on real compounds whose atoms and molecules we do not see. The fact that the mass ratios observed with real compounds correspond to those which we produced with fasteners and rings contributes to our confidence in the atomic model.

The second idea is the recognition of the law of multiple proportions as an extension of the law of constant proportions; the law of constant proportions presented in Chap. 6 is not "wrong" but is limited to single compounds. The law of multiple proportions thus becomes a special case of the more general law of constant proportions.

---

*If you keep some of the chlorides for future use, store them in airtight bottles to keep them from absorbing moisture and, therefore, from being difficult to dry out. Also, if damp, the copper (I) chloride will oxidize to the copper (II) chloride.

MOLECULES 8.9

This section serves primarily to show that only for gases is it always meaningful to speak of molecules; in solids and liquids atoms may not form well-defined clusters. We only hint that the compressibility of solids and liquids may be an indication of the existence of clusters of atoms and thus suggests the existence of empty spaces between them. Gases are much more compressible than liquids; liquids with a molecular structure are more compressible than those without; and crystals are least compressible of all.

Students may suggest that the compressibility of gases can be explained by particles with spongelike properties; the compressed "sponge" is the particle of the solid, and the expanded sponge is the particle of the gas. Using the evidence discovered so far in this course, they cannot choose between these two models unless they take into account the mobility of gases as we shall do in Chap. 9. The point to emphasize at this stage is that the atomic model for gases and solids as described in this section is consistent with the observations on compressibility.

RADIOACTIVE ELEMENTS AND THE ATOMIC MODEL 8.10

Students should never get the idea that the theories, models, or laws of science are forever fixed and not subject to modification. The modification of our atomic model to include the observations of the radioactive process is one of several examples of the modification of a model in the course.

The PSSC film, *Definite and Multiple Proportions*, can be shown at the end of the chapter as an additional example of multiple proportions and as a review of the chapter.

CHAPTER 8—ANSWERS TO PROBLEMS

Sec.	Easy	Medium	Hard	Class Discussion	Home or Lab
1	1†, 2	20		20	2
2		3, 21			
3		4			
4	5, 7	6, 8		6	
5	9†, 22	10		22	
6		11			
7		12, 14	13		
8		15†, 16†, 17, 23	26		
9		18, 19	24, 25 27, 28	18, 19, 28	

1† In addition to enabling us to summarize and account for the facts we have obtained from observations and experiment, what should a good model do for us?

It must suggest at least one new experiment and correctly predict the results.

2 You have two eggs in the refrigerator. One is raw and the other is hard boiled, but you do not know which is which. You spin one egg and it spins easily. You spin the other and it comes to rest after a very few rotations.

a) Suggest a model that will account for the different behavior of the two eggs.
b) On the basis of your model, what do you predict about the spinning of the two eggs if you boil both of them for 10 minutes?
c) Check your prediction at home.

a) One egg is mostly liquid inside; this egg after being spun comes to rest quickly. The other is solid inside; this one spins easily.
b) Both eggs should spin equally well.

3 a) In investigating the black box you did certain things to it that you could easily undo. Give some examples.
b) In investigating the characteristic properties of a substance you can also do things that can easily be undone. Give some examples.
c) What did you do to the black box that you could not undo?
d) Does dissolving zinc in hydrochloric acid resemble any of the kinds of tests you mentioned in part (a) or part (c)?

a) You shook the box, tilted it, turned it around, and turned it over.
b) Boiling a liquid and condensing the vapor; melting a solid and then letting it solidify; dissolving a solid in a liquid and then evaporating the liquid.
c) You pulled one or more of the rods from the box.
d) It resembles part (c). (Although zinc can be recovered from zinc chloride, this is not done in the IPS laboratory.)

4 Suppose that *M* atoms of mercury combine with *N* atoms of oxygen to form mercury oxide.
a) What total number of atoms would you expect there to be in the sample of mercury oxide produced?
b) If the mercury oxide produced is then decomposed by heating to form gaseous mercury and oxygen, how many atoms of mercury and how many atoms of oxygen would you expect to find?
c) Would your answers to (b) be different if you condensed the samples of gas to liquid mercury and liquid oxygen?

a) The sample of mercury oxide would contain M + N atoms since, according to the atomic model, the atoms do not change in number when they react to form a compound.
b) *M* atoms of mercury reacted with *N* atoms of oxygen to form mercury oxide. If the atoms are essentially unchanged as a result of the reaction, when the mercury oxide is decomposed you would expect to find the same number of atoms: *M* atoms of mercury and *N* atoms of oxygen.
c) No, we expect the number of atoms to remain the same.

5 Does the experiment with rubber rings and paper fasteners give you any evidence for the shape of an atom?

The choice of rings and fasteners in the experiment has nothing to do with their shape or even their ability to fasten together. The difference in masses of the rings and fasteners is the only property utilized in this experiment. Any other set of objects with different masses would work just as well.

6 In the synthesis-of-water experiment described in Sec. 6.3, different amounts of hydrogen were mixed with the same amount of oxygen, and then the mixture was ignited by a spark. In the first case there was some oxygen left over; in the second case nearly all the oxygen and hydrogen reacted; and in the third case hydrogen was left over. Describe the results of these experiments in terms of the atomic model of matter.

The atoms of hydrogen and oxygen combined in only one ratio to form water. Therefore, in the first case all the atoms of hydrogen combined with the atoms of oxygen in a 2:1 ratio by volume, and there was an excess of oxygen atoms left uncombined, if the reaction went to completion. Similarly, in the last case, after two volumes of hydrogen atoms combined with one volume of oxygen atoms, there was a number of hydrogen atoms remaining uncombined. Only in the case where the ratio of hydrogen volume to oxygen volume was 2 did nearly all the hydrogen atoms combine with nearly all the oxygen atoms, with practically nothing left over.

(This experiment by itself does not tell us that in the synthesis of water two atoms of hydrogen combine with one atom of oxygen!)

7   Suppose that element A can form a compound with element B but not with element C. Element B can form a compound with C. How would you choose fasteners, rings, and washers to represent these elements?

You could choose rings to represent A, fasteners to represent B, and washers to represent C.

8   Describe the synthesis of zinc chloride (Expt. 6.4) in terms of fasteners, rings, and washers. (Hint: Hydrochloric acid contains hydrogen and chlorine.)

You could choose rings to represent hydrogen, washers to represent zinc, and fasteners to represent chlorine; then hydrogen chloride would be represented by a ring on a fastener. The synthesis of zinc chloride would be made by removing the ring from the fastener and putting a washer on the fastener instead. (Don't penalize students who assume that hydrochloric acid is HCl_2 or H_2Cl, etc. rather than HCl!)

9† The ratio of the mass of lead to the mass of oxygen in an oxide of lead is 13. How many grams of lead will combine with 100 g of oxygen?

1,300 g

10 You can make several "compounds" by gluing pennies (Pe) and nickels (Ni) together. The ratio of the mass of Ni to Pe in the compound NiPe is 1.6. What will be the ratio of the mass of Ni to Pe in the compound (a) Ni_2Pe, (b) Ni_3Pe, (c) $NiPe_2$?

This ratio tells us that each Ni atom has 1.6 times the mass of one Pe atom. Therefore,

a) in Ni_2Pe, Mass Ni/Mass Pe = 2 × 1.6 = 3.2
b) in Ni_3Pe, Mass Ni/Mass Pe = 3 × 1.6 = 4.8
c) in $NiPe_2$, Mass Ni/Mass Pe = $\frac{1.6}{2}$ = 0.8

11 Two samples of two compounds containing only nuts and bolts were decomposed. A mass of 100 g of each sample yielded the following data.

Compound sample	Mass of bolts (g)	Mass of nuts (g)
1	80	20
2	67	33

a) What mass of nuts would combine with 100 g of bolts in sample 1?
b) What mass of nuts would combine with 100 g of bolts in sample 2?
c) What is the ratio of the mass of nuts in sample 2 to the mass of nuts in sample 1 that would combine with 100 g of bolts?

a) In sample 1, (100/80) × 20 g = <u>25 g</u> nuts will combine with 100 g of bolts.
b) In sample 2, (100/67) × 33 = <u>50 g</u> of nuts will combine with 100 g of bolts in sample 2.
c) The ratio is 50/25 = <u>2.</u>

12 In the experiment you have just done, how would the value for the mass of chlorine combining with 100 g of copper in the brown powder have been altered if this powder had not been thoroughly dried before massing?

If the brown powder had not been thoroughly dried before massing, the moisture would add to its mass. Subtraction of the mass of dried copper from that of the brown powder would give too high a mass of chlorine. This would result in too high a value for the mass of chlorine combining with 100 g of copper.

13 In the case of the brown powder, suppose that some copper had been left in the beaker after the wet copper was transferred to the watch glass. How would this error have affected the results of the experiment?

The mass of copper transferred to the watch glass would have been too small, and the calculated mass of chlorine in the brown powder would have been too large. Since the mass of chlorine that combines with 100 g of copper is given by the expression $100 \times \frac{\text{mass of chlorine}}{\text{mass of copper}}$, the calculated amount of chlorine combining with 100 g of copper would have been too large.

174   The Atomic Model of Matter

14 A student repeated the experiment with the brown copper chloride several times, and each time he used the same strip of aluminum. In his last determination he found that there was no aluminum left to recover. How would this affect his value for the mass of chlorine combined with 100 g of copper in the last determination?

If all the aluminum dissolved in an experiment of this kind, it is unlikely that all the copper would have been displaced from the chloride. The mass of the copper would be too small, and consequently the determination of the mass of chlorine combined with 100 g of copper would be too large.
   (If the solution remains colored, this would suggest to the student that he should use additional aluminum to complete the experiment.)

15† What is the ratio of the mass of chlorine that combines with 100 g of phosphorus in phosphorus trichloride to the mass of chlorine that combines with 100 g of phosphorus in diphosphorus tetrachloride (Table 8.3)?

   $1.5 = 3/2$

16† From the data given below for compounds of lead and oxygen, calculate the ratios in the last column. Do these ratios agree with the law of multiple proportions?

Compound	Formula	Mass M of oxygen combined with 100 g of lead (g)	Mass ratio
Lead (I) oxide	Pb_2O	$M_1 = 3.86$	—
Lead (II) oxide	PbO	$M_2 = 7.72$	$\dfrac{M_2}{M_1}$
Lead (III) oxide	Pb_2O_3	$M_3 = 11.58$	$\dfrac{M_3}{M_1}$
Lead (IV) oxide	PbO_2	$M_4 = 15.44$	$\dfrac{M_4}{M_1}$

   $M_2/M_1 = 2.00$ $M_3/M_1 = 3.00$ $M_4/M_1 = 4.00$
   Yes.

17 Two compounds containing only carbon and oxygen are decomposed. A mass of 100 g of compound I contains 43 g of carbon, and 100 g of compound II contains 27 g of carbon.
   a) What is the ratio of the mass of carbon to the mass of oxygen for each compound?
   b) If compound II has the formula CO_2, what is a possible formula for compound I?

a) Compound I contains 43 g of carbon and (100 − 43) g = 57 g of oxygen, giving a ratio for the mass of carbon to the mass of oxygen of 43 g/57 g = 0.75.

Compound II contains 27 g of carbon and (100 − 27) g = 73 g of oxygen, giving the ratio for the mass of carbon to the mass of oxygen of 27 g/73 g = 0.37.

b) From (a) it can be seen that 75 g of carbon will combine with 100 g of oxygen to form compound I, and from (b) it can be seen that 37 g of carbon will combine with 100 g of oxygen to form compound II. Therefore, compound I has 75/37 = 2 times as much carbon as compound II for the same mass of oxygen. If compound II has a formula of CO_2, compound I could be CO.

Students may give the answer C_2O_2. You can point out that this can be more simply expressed as CO. The formula C_2O_2 has meaning only in terms of molecules, and molecular formulas such as H_2 and O_2 are not discussed until later in the chapter.

18 **Why is it more difficult to speak in terms of molecules in a solid than in a gas?**

The incompressibility of solids indicates that the atoms are very close together. This makes it difficult to think of clusters of atoms where the distance between the centers of atoms in adjacent clusters is greater than the distance between the centers of atoms in a single cluster.

19 **What experimental evidence indicates that some solids do consist of a collection of molecules?**

The fact that some solids (and liquids) can be compressed significantly more than others indicates that there are clusters or groups of atoms with small spaces in between. When the substance is compressed, the spaces between these clusters or molecules are reduced.

20 **Think again of the sealed tin can referred to in Sec. 8.1. You are not allowed to pierce the can or break it open. What would you predict about its behavior if (a) you lowered the temperature sufficiently or (b) you raised the temperature sufficiently?**

a) If the contents of the can are like a liquid and the temperature is lowered far enough, we should predict that the liquid would freeze and that no more "sloshing" would take place.

b) We should predict that the liquid would vaporize and that the can might blow apart from the increase in pressure. If the quantity of liquid were not too large, we should predict that it would all turn to vapor, without damaging the can, and that no more "sloshing" would take place.

176 The Atomic Model of Matter

21 While doing Expt. 8.2, A Black Box, why were you not permitted to open the box and look inside?

The black box experiment was conducted in such a way as to illustrate some of the problems involved in constructing a model, and in particular, an atomic model of matter. Since you cannot see "inside" matter, you were asked not to open the box.

22 Which of the following are *parts* of our atomic model of matter? Which may be used to *illustrate* the model?
 a) Matter is made up of very tiny particles, much too small to be seen.
 b) A paint sprayer will deposit more particles the longer it is operated.
 c) Atoms of the same element are all alike.
 d) Two rubber rings combine with one fastener to form FsR_2.
 e) Atoms of two different elements may combine to form compounds.
 f) Atoms of two different elements may combine in different ratios to form different compounds.
 g) Marbles can be stacked in a box to represent a solid.

(*a*), (*c*), (*e*), and (*f*) are part of the model, (*b*), (*d*), and (*g*) are only analogies used to illustrate the model.

23 A mass of 16 g of oxygen combines with 63.5 g of copper to form CuO.
 a) What is the ratio of the mass of copper to the mass of oxygen?
 b) What is the ratio of the mass of copper to the mass of oxygen in Cu_2O?

a) The ratio of the mass of copper to the mass of oxygen in CuO is 63.5 g/16 g = <u>4.0</u>.
b) Cu_2O has twice as much copper for the same mass of oxygen, giving a ratio of copper to oxygen of (2 × 63.5 g)/16 g = <u>7.9</u>.

24 In your own words state the law of multiple proportions.

The answers the students will give will vary, but all should contain the following ideas.
a) The same two elements must form at least two different compounds in order for the law to be applicable.
b) One must consider a mass of each compound such that the same mass of one of the elements is contained in all compounds.
c) Then the mass of the second element in any one of the compounds will form a ratio of small whole numbers when combined with the mass of the second element in any one of the other compounds.

25 Carbon can combine with chlorine to form three different compounds, CCl$_4$, C$_2$Cl$_4$, and C$_2$Cl$_6$. If the ratio of the mass of carbon to chlorine in CCl$_4$ is 0.286, what is the ratio of the mass of carbon to that of chlorine in the other two compounds?

In C$_2$Cl$_4$ there are half as many chlorine atoms combined with one carbon atom as there are in CCl$_4$, and the ratio $\frac{\text{mass of carbon}}{\text{mass of chlorine}}$ is $\frac{0.286}{1/2} = 0.572$. In C$_2Cl_6$ there are ¾ as many chlorine atoms combined with one carbon atom as there are in CCl$_4$, and the ratio $\frac{\text{mass of carbon}}{\text{mass of chlorine}}$ is $\frac{0.286}{3/4} = 0.381$.

26 Suppose that 6×10^6 atoms of hydrogen combine with 2×10^6 atoms of nitrogen and form 2×10^6 molecules of ammonia. How many atoms of hydrogen and how many atoms of nitrogen would there be in each molecule of ammonia?

To find the number of hydrogen atoms and nitrogen atoms in each molecule of ammonia, we divide the number of hydrogen atoms and the number of nitrogen atoms by the number of molecules. Number of hydrogen atoms per molecule:

$$\frac{6 \times 10^6}{2 \times 10^6} = \underline{3}$$

Number of nitrogen atoms per molecule:

$$\frac{2 \times 10^6}{2 \times 10^6} = \underline{1}$$

Therefore, the formula for ammonia would by NH$_3$.

27 A student suggested the following formulas for the two chlorides of copper.

Brown Chloride	Green Chloride
CuCl$_2$	CuCl
CuCl	Cu$_2$Cl
Cu$_3$Cl$_2$	CuCl
Cu$_3$Cl$_4$	Cu$_3$Cl$_2$

a) Which pair or pairs are possible formulas for these compounds? pounds?
b) Can you decide which of the pairs gives the correct formulas for the chlorides of copper?

a) The results of Expt. 8.7 indicated that the number of chlorine atoms that combine with one copper atom in the brown compound is twice the number that combine with one copper atom in the green compound. All the above pairs of formulas for the brown and green compounds agree with the results of the experiment, except for the pair Cu_3Cl_2–$CuCl$, in which the green compound has 3/2 times as much chlorine as the brown.

b) Without knowing the relative masses of the individual atoms, we cannot decide which of the pairs gives the correct formulas for the two compounds. The determinations of correct simplest formulas from the relative masses of the atoms will be discussed in Sec. 10.7.

28 How does the atomic model of matter enable us to account for (a) the law of constant proportions, (b) the law of multiple proportions, and (c) the law of conservation of mass?

a) We assume that the ratio of the numbers of atoms of two elements combined in a given compound is always the same. Therefore, the ratio of the masses of these elements will be constant since, according to the atomic model, all atoms of a given element have the same mass.

b) If two elements can react to form more than one compound, we assume that the ratio of the numbers of atoms of the two elements is different in the different compounds. According to the model, all atoms of a given element have the same mass. Therefore, if there are twice or three times, etc., as many atoms of one element combined with each atom of the other element in one compound as there are in another compound, the ratio of the masses of one element combined with a fixed mass of the other element in the two compounds will be 2, 3, etc.

A pair of compounds in which there are three atoms of element A for each atom of element B in one compound and four atoms of A for every atom of B in the other compound leads to the mass ratio 3/4.

c) According to the model, the atoms remain unchanged in number and mass when a compound is formed. Therefore, when two elements combine to form a compound, the total mass before the reaction will equal the total mass after the reaction.

Molecular Motion

9

Overview of the Chapter

Until now in this course we have considered molecules as static clusters of atoms. Even when we discussed reactions in which compounds were formed, we limited our attention to the arrangement of the atoms before and after the reaction took place. But it is quite clear that no reaction can take place without the atoms moving. In this chapter we add another dimension to the atomic model of matter, namely that of motion. This motion is most apparent in gases, and that is one of the reasons for starting with gases. Another reason is that in some respects all gases have similar properties that can be related to molecular motion alone, without reference to the nature of the molecules themselves.

By using a mechanical analogue to a real gas, and some simple reasoning, we can see how molecular motion can account for the elastic properties of gases. On the basis of the ideas gained from these considerations, we can make a prediction about the relation between the volume and the pressure of a gas. This prediction is verified by an experiment described in the text, and we are able to state Boyle's law as a consequence. However, students are cautioned that the similarity among gases disappears gradually as the condensation point is approached; Boyle's law, like other laws, is valid only over a limited range of conditions.

The mechanical analogue to a real gas also guides us into a qualitative discussion of the relation between temperature and molecular speeds in gases. The thermal expansions of gases, liquids, and solids are investigated; the latter two are found to be characteristic of particular substances, whereas the thermal expansion of gases are all the same (as long as the pressure is not too high).

Most of the remainder of the chapter is devoted to a qualitative discussion of this relation between temperature and molecular speed and molecular motion in liquids and solids.

The suggested schedule for this chapter is:

Section 1 (one demonstration, probs. 1-2, 18-20)   1 period
Section 2 (one demonstration, probs. 3-6, 21-22)   1 period

180 Molecular Motion

Sections 3-5 (one demonstration, probs. 7-11, 23-24) 2 periods
Section 6 (one demonstration, probs. 12-13, 25-28) 1 period
Sections 7-9 (two film loops, one expt., probs. 14-16, 29-35) 4 periods
Sections 10-11 (one experiment, probs. 17, 36-38) 2 periods
Film: "Crystals" 1 period
Total 12 periods

9.1 MOLECULAR MOTION AND DIFFUSION

Although the details and the discussion of the demonstration are complete in the text, your students will have a far better understanding of the process if they can actually see it take place. If you do not have the bromine tubes, show the IPS film loop "Molecular Motion and Diffusion (with Bromine)."

If you have the bromine tubes and access to Dry Ice, the following suggestions will be helpful.

The demonstration with the bromine tubes can easily be performed during a class period. The dark gas is best seen if the tubes are placed against a white background, as was done in making the photographs in Figs. 9.1 and 9.2. The photograph in Fig. I shows a suitable way to mount the tubes, using a pegboard, large clamps, and a piece of heavy white cardboard.

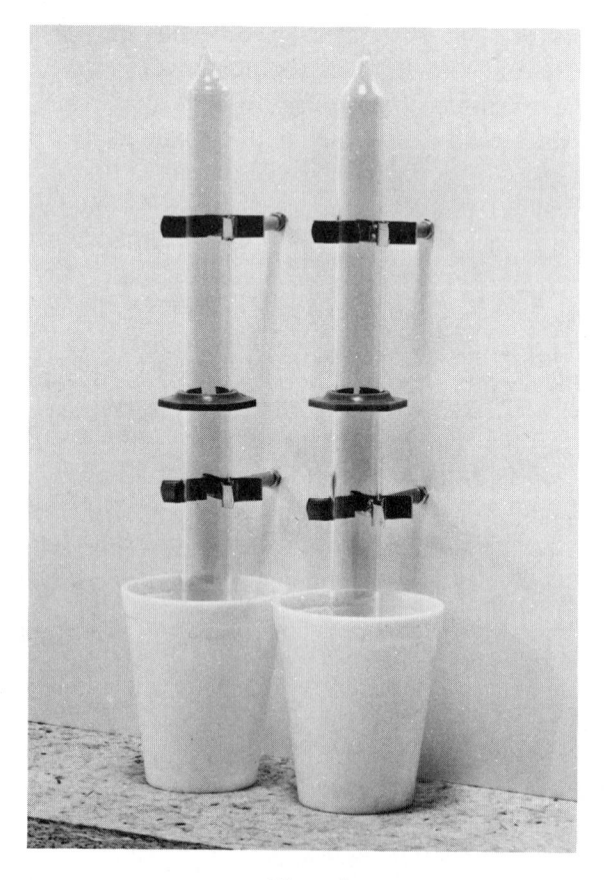

Figure I

To make the cooling mixture, crush some Dry Ice, put it in a Styrofoam cup, and add a little burner fuel. To have all the bromine solidify at one place, dip only about half a centimeter of the bottom of the tube in the cooling mixture.

If Dry Ice is not available, a carbon dioxide fire extinguisher can be used (as described in the *Teacher's Guide* for Sec. 7.9), to supply solid carbon dioxide for the experiment. The solidified carbon dioxide can be scraped off with a ruler edge into a Styrofoam cup. One 3- or 4-second operation of the extinguisher should give enough Dry Ice for one cup.

The bromine will condense and freeze out in the bottom of the tube, and the dark-brown color will disappear. The bromine in the tube that contains air will require a longer time to condense than the bromine in the tube that contains no air. In the tube containing air, most of the bromine will solidify in 10 minutes in the cooling bath. However, a little bromine will still remain in the upper part of the tube, even after several hours of cooling.

Freeze out the bromine before class, so that when you start the demonstration, the bromine gas is solidified in the bottoms of both tubes. Remove the tubes from the cooling mixture and display them against a white background to your students. If the tubes have been cooled sufficiently, there is almost no color in either at first. However, the evacuated tube will begin almost immediately to turn light brown uniformly throughout its length. Soon the tube containing air will begin to turn light brown at the bottom, and the color will slowly diffuse up the tube.

If the tubes are left undisturbed long enough (20 to 30 minutes) at room temperature, the bromine will be distributed evenly throughout both tubes, and again no difference can be detected. When the tubes are cooled once more with Dry Ice, students can see how differently the bromine gas behaves while it is condensing in air and in a vacuum.

Although this demonstration shows that gas molecules move about by themselves, it does not give much information about their speed—only that when they move unimpeded (in the tube without air), they move the length of the tube faster than the eye can detect.

Bromine melts at $-7.2°$ and boils at $58.8°C$, which is above room temperature. At room temperature, it is very volatile, and the small amount of bromine in the tubes evaporates completely, so that no liquid bromine is visible in the tubes. Bromine is very corrosive to the skin and lung tissue. If a tube should break, do not touch the liquid or breathe its vapor. There is so little bromine in the tubes, however, that in case of accident, opening the windows will be sufficient to eliminate any danger.

Apparatus and Materials

Set of bromine tubes
2 Styrofoam cups
Dry Ice
Burner fuel

Pegboard
4 Large clamps
White paper or cardboard

182 Molecular Motion

Advance Planning

At this time you may want to start the copper sulfate demonstration, described in Sec. 9.10 in order to compare the rate of diffusion in gases and in liquids.

9.2 DENSITY AND PRESSURE OF A GAS

The sphere-gas machine discussed in this section is used to help your students visualize what a gas may be like. The spheres in the machine are an analogue of a real gas, and in many ways they behave much like real gas particles confined in a cylinder. However, just as we had to be careful in Chap. 8 to distinguish between the atomic model for compound formation and its illustration in terms of fasteners and rings, so must we be careful to distinguish between the molecular model of the pressure of a gas and its illustration with steel spheres. In Chap. 8 we ignored the shape and size of the fasteners and rings and concentrated on their ability to combine in definite ratios. Here we ignore the fact that steel spheres do not bounce around by themselves, that they are many times larger (10^7) in diameter than gas molecules, and that they move much more slowly than gas molecules. We note that when they are in motion, they exert a pressure on the walls and piston of the container; we study, at least qualitatively, how this pressure varies with the number of spheres and with the average speed. Therefore, when you discuss the sphere-gas machine, point out to your students that it will not reproduce exactly the behavior of a gas in a cylinder but will help them to visualize some of the effects of molecular motion.

Show the IPS film loop "The Sphere-Gas Machine (an Analogue)." Your students will have a much better idea of the nature of gases when they see the demonstration carried out; it will be easier for them to see what happens than it would be if you do the demonstration yourself. However, if you have the machine and would like to have some of your students experiment with it directly, here are some additional points on its use.

The demonstration in this section deals with the effect that the number of spheres in the cylinder has on the volume and the pressure. When it comes to demonstrating the properties of gases quantitatively, the machine is a poor analogue of the gas model. It is not worthwhile to try to get meaningful quantitative data from it. Only over a very narrow range of pressure, motor speed, and number of spheres (different for every machine) will the volume approximately double when the number of spheres is doubled. However, the apparatus is very useful to demonstrate qualitatively in a graphic way that the volume increases when the number of spheres is increased, and that the pressure must be increased if we want to get back to the original volume.

To remove the spheres from the cylinder, lower a small magnet into it on a thread.

The apparatus runs with a 6-volt battery, and the pressure can be changed by stringing rubber grommets on the piston guide or by pushing

Molecular Motion   183

down on the piston-guide rod. (Let your students do this so they can feel the increase in pressure of the spheres as they decrease the volume.) The grommets can be held firmly in place on the guide rod by means of a small cork that fits tightly on the guide rod.

Apparatus and Materials

Sphere-gas machine   Pegboard
Six-volt battery   2 Large clamps
Connecting wires   Small magnet on thread

9.3 A PREDICTION ABOUT THE RELATION BETWEEN VOLUME AND PRESSURE OF GASES

Here we use the ideas suggested in the previous section to make a prediction about the way in which the volume of a gas is related to its pressure. It will be advantageous to read the entire section in class.

9.4 THE COMPRESSIBILITY OF GASES

The description of the experiment on compressibility of gases given in this section will present difficulties for students who have not seen the demonstration. The section is intended to serve as a reminder of what the students saw. Therefore, we suggest that either you yourself demonstrate the compressibility or that you have some of your students do so. Although the procedure outlined is simple, some advance practice will help you to get reliable readings.

It is possible that some of your students may have difficulty in understanding why it is necessary to add 2.4 bricks to all pressure readings to give a graph (Fig. 9.15) in agreement with the graph in Fig. 9.11. You can discuss with them what would happen if the apparatus shown in Fig. 9.12 were placed in a room with no air in it. The platform would rise, and bricks equivalent to the air pressure would have to be placed on it to bring the volume back down to what it was in a room containing air at atmospheric pressure.

You can also remind your students about how the volume of the "sphere gas" in the machine was reduced to half in the experiment described in Sec. 9.2 by adding a disk that doubled the mass of the piston. It was only when the students compared the *total* masses pressing down on the "gas" before and after increasing the mass that the ratio of two-to-one for the masses was obtained. (Atmospheric air pressure pushing down on the piston did not come into play in the sphere-gas machine because both sides of the piston were open to the atmosphere.)

9.5 BEHAVIOR OF GASES AT HIGH PRESSURES

The behavior of gases at high pressures provides another example of the limitations of the laws of nature and the danger of large extrapolations. This section is worth reading in class.

We use the term "characteristic property" in this section for properties such as density, or compressibility at high pressures, which are independent of mass and vary from substance to substance, and which are useful in identifying the substance. In discussing the section with your students, it may be helpful to refer to these properties as "distinguishing" characteristic properties. For instance, compressibility is a characteristic property of all gases at low pressures, since it is independent of the amount present; but it becomes a *distinguishing* characteristic property only at high pressures.

9.6 TEMPERATURE AND MOLECULAR SPEED

Again, use the machine to amplify the text by showing the effect of faster- and slower-moving molecules. Two 6-volt batteries can be connected in series to run the machine at a higher speed for a few minutes. A resistor may be put in series with a single battery for slower speeds. A nichrome heating element, of the kind used in a hot plate, makes a good resistor for this experiment if you do not have a suitable rheostat. By varying the length of nichrome wire in the circuit, the speed of the motor can be changed. The results will again be only qualitative, showing that the volume increases with the speed of the motor.

A fire syringe, like the one shown in Fig. 9.18, can be used to demonstrate that the temperature of a gas rises as it is compressed. It generally requires some practice before the cotton can be ignited consistently. Place a very small piece of cotton or Kleenex *loosely* in the bottom of the ignition tube. You may have to push the piece to the bottom with a small rod. Lubricate the rubber O-ring with any rubber lubricant or stop-cock grease. Place the piston in the end of the glass tube, insert the glass tube into the metal cylinder, and support the end of the metal cylinder firmly on the desk. The metal cylinder is for protection against broken glass in case the tube should burst; do not try the demonstration without it. When a quick, hard push is given to the piston, a flash of light can be observed as the cotton ignites.

9.7 THERMAL EXPANSION OF GASES

The material in this section should be treated as a class discussion of the experiment and the data presented in the IPS film loop "How Does the Thermal Expansion of Gases Compare?" The fact that gases expand at the same rate is important for our model of matter. Thermal expansion, unlike many characteristic properties, does *not* distinguish between gases.

Do not attempt to perform the experiment described in the loop as a quantitative demonstration. The experiment is very time-consuming. However, it is worthwhile to demonstrate the thermal expansion of gases qualitatively in order to make it more vivid.

Insert a straight piece of glass tubing into a No. 4 one-hole stopper. In order to get a small plug of water in the tubing, dip the stopper end of the tubing into some water. Insert the stopper into a large test tube, taking care not to dislodge the water plug. The water plug will rise in the tube as you warm the gas simply by holding the tube in your hand. This shows the large expansion of gas for a small temperature change when compared with the expansion of liquids discussed in Sec. 9.8.

THERMAL EXPANSION OF LIQUIDS 9.8

Show the IPS film loop "How Does the Thermal Expansion of Liquids Compare?"

Here students see equal volumes of three liquids heated equally by having the tubes containing them immersed in the same water bath.

Since the expansion of the liquids is small, amplification must be used. The change in volume is amplified by measuring the liquid rise in the narrow glass tube rather than in the large test tube. This notion of amplifying a small change in volume is utilized in any mercury or alcohol thermometer.

If you have some ambitious students who would like to do the experiment for themselves, here are some helpful details on setting up and conducting the experiment.

Use three small clamps to hold the three large test tubes in a 600-ml water bath set on a burner stand. With this arrangement the level of liquid in the three tubes can be made the same at the start of the experiment in the following way: Insert a piece of glass tubing about 15 cm long (6 mm outside diameter tubing is best) into one hole of a No. 4 two-hole stopper. Leave the other hole open. Pour liquid into the test tube so that the liquid rises in both holes. Excess liquid may flow from the free hole until the liquid is level with the top surface of the stopper in both holes. Push a glass stirring rod into the free hole. You can now adjust the level of the liquid in the glass tubing by moving the rod up and down.

The stopper used for the glycerin will have a tendency to pop out if there is a considerable amount of the liquid between the stopper and the wall of the test tube. You can avoid this by taking care not to put too much glycerin in the test tube at the start. Wiping away excess liquid with a paper towel as it oozes from the free hole will help also. To obtain the best results for the glycerin, use a fresh sample that has not absorbed water by being exposed to the air for any length of time.

Air bubbles forming under the stopper can be coaxed from the tube by use of a straightened paper clip inserted into the free hole. Small air bubbles will not affect the results appreciably. The demonstration requires about 15 minutes if two alcohol burners are used. Be careful not to allow the burner fuel to overflow.

Apparatus and Materials

Pegboard
3 Small clamps
3 Large test tubes
3 No. 4 two-hole rubber stoppers
3 Straight pieces of glass tubing
    (6 mm in diameter, about
    15 cm long)

4 Glass stirring rods
    1000-ml beaker (a 600-ml beaker
        can be used if necessary)
    Burner stand
2 Alcohol burners
    Water
    Glycerin, burner fuel

186 Molecular Motion

9.9 EXPERIMENT: THERMAL EXPANSION OF SOLIDS

A discussion of the coefficients of linear expansion has been left out on purpose. It is too time-consuming and adds little to the development of the basic ideas of the course. Emphasize in the post-lab discussion that the expansion of the various substances was compared for samples of equal length and over the same temperature range.

The Experiment

One end of the tube must be anchored to ensure that the expansion takes place in only one direction. A band of masking tape wrapped around the tube will make the clothespin hold it more firmly. The moving end of the tube is pressed firmly down on the indicator pin by a rubber band fastened to a thumbtack stuck into the bottom of the pegboard. (See Fig. 9.22(b).) This arrangement is worth discussing and demonstrating in the pre-lab discussion.

This is another experiment in which the effect to be measured is so small as to require an amplifying device. Considerable care must be taken to see that there is no slippage in the apparatus. Also, the indicator dial must be free to rotate without rubbing. A fiberboard slide under the dial end of the tube provides a smooth surface on which the needle can rotate.

Care should be taken that the rubber band holding down the exhaust end of the tube is tight enough to cause the tube to rotate the needle without slippage, but not so tight that the motion of the tube is restricted, as this would result in the tube expanding in the opposite direction, through the clothespin.

The indicator dial should be glued or taped to the large "T" pin so that it does not rotate with respect to the pin. The dial can be set at zero by lining up the zero-degree mark with the edge of the pegboard base. The pegboard must be placed near the edge of the table or on blocks to allow the dial to rotate without striking the table.

The rubber tubing must be connected to the metal tube and the pointer must be set at zero before the water starts to boil in the test tube. If this is not done, students will burn themselves, and the pointer will not be at zero before the tube starts to heat up. The steam hose should be short enough that it always slants down toward the clothespin. If it drops below the tube, a water trap will form, partially blocking the flow of steam.

The test tube should be a large one and no more than half full of water. Otherwise, hot water rather than steam may be forced into the plastic tubing caused by the violence of the boiling water. If available, a small flask may be used instead of the test tube. This will obviate the necessity of replacing water that has been boiled away. In any case, a few boiling chips should be added to the water.

The short rubber tube on the exhaust end of the tube being tested should not fit tightly around *both* the tube and the thermometer. The bulb of the thermometer must be inserted well into the rubber tube for best the thermometer must be inserted well into the rubber tube for best results. If the thermometer fits too tightly, there will be a pressure buildup with consequent popping of the rubber stopper in the test tube and danger of scalding by hot water. The thermometer can be taped to the table to prevent it from rolling onto the floor.

To reduce the chance of mishaps, you should demonstrate the proper use of the apparatus, calling particular attention to the following points. Make sure that the tube does not roll on the tapered portion of the needle. Do not push the thermometer too far into the rubber tubing. Read the amount of expansion before making a measurement of temperature, lest the tube be disturbed when making the temperature measurement. Remove the rubber band at the end before removing the glass tube from the board to prevent breakage.

Some students may wish to calculate the actual change in length as an extension of this experiment. To do this, they must measure the diameter of the rolling needle and calculate the amplification of the needle and indicator. Figure II shows that as the needle is rolled by the expanding tube through 360°, the center of the needle progresses a distance equal to one circumference with respect to a point on the table; therefore, the tube expands a distance equal to twice the circumference of the needle. For one degree of rotation of the indicator, therefore, the tube expands a distance

$$\frac{2\pi D}{360} = \frac{\pi D}{180}$$

where D is the diameter of the needle. For α degrees, the change in length is $\pi D \alpha/180$.

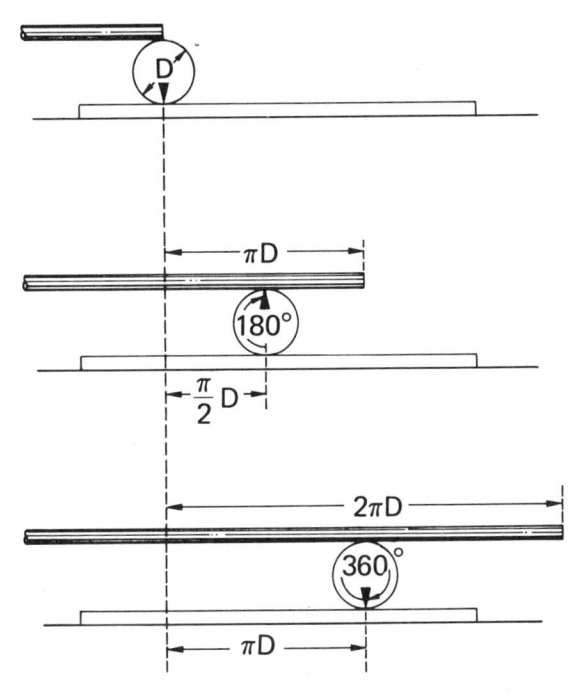

Figure II

Sample Data

The pin to which the dial was attached had a diameter of 0.12 cm, and the length of the tube between the clothespin and the dial pin was 43 cm. The indicator was set at zero when the tubes were at room temperature (28°C).

	Trial 1		Trial 2	
Tube	Temp. (°C)	Indicator Reading (degrees)	Temp. (°C)	Indicator Reading (degrees)
Copper	98	28	97	29
Aluminum	97	36	98	36
Glass	98	6	97	6

Answers to Questions

All the tubes were heated through the same temperature range (within about 1 percent).

As the histogram in Fig. III indicates, the expansion of tubes of equal length heated through the same temperature change depends on the substance of which the tubes are made. Therefore, this expansion is a characteristic property of solids.

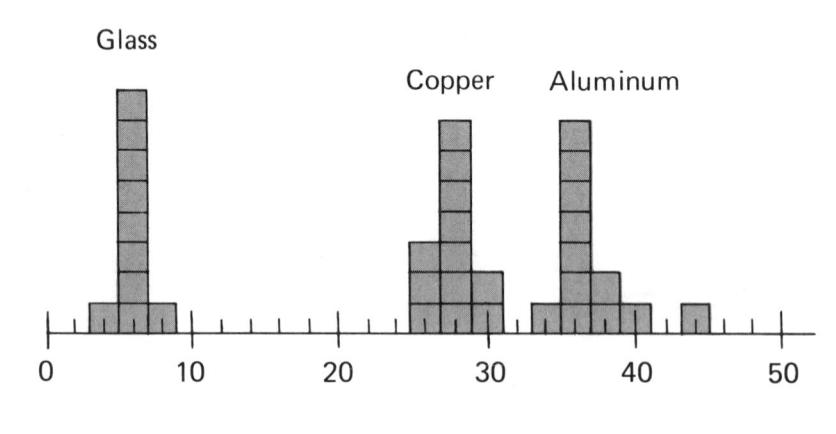

Figure III

Apparatus and Materials

Pegboard
Clothespin with screw, wing nut, and washer
Large needle, cardboard dial
1-hole stopper
Right-angle glass bend
12-in. length of rubber or plastic tubing
1-in. length of rubber or plastic tubing
  (inside diameter greater than the diameter of the thermometer)
Fiberboard (1 in. × 2 in.)
Set of three thin-walled expansion tubes, 0.60 cm diameter
2 Alcohol burners
Boiling chips
Large test tube
Clamp
Thumbtacks
Rubber bands
Centimeter rule
Water
Watch glass
Paper towel
Quick-drying glue
Masking tape
Thermometer
Safety glasses

MOLECULAR MOTION IN LIQUIDS AND SOLIDS 9.10

The molecular diffusion occurring between water and a copper sulfate solution provides a good project for two students to demonstrate to the entire class. The problem is to place in a 250-cm³ graduated cylinder a layer of 100 cm³ of water on top of 100 cm³ of almost saturated copper sulfate solution so that the two layers are sharply separated. This can be done by first pouring the water into the cylinder and then slowly adding the copper sulfate at the bottom of the cylinder through a long glass tube. The tube should be softened in a hot flame and drawn to make a fine opening like an eyedropper at the lower end, or you can use a long glass tube connected to an eyedropper by a short length of rubber or plastic tubing. This should be connected at the upper end to a short-stem funnel by means of a rubber tube. A pinch clamp on the rubber tube allows careful control of the flow of liquid.

After adding the water to the graduated cylinder, fill the funnel with a solution made by dissolving 40 g of blue copper sulfate ($CuSO_4 \cdot 5H_2O$) in 150 cm³ of water. The addition of 1 cm³ of concentrated sulfuric acid to the solution will prevent brown copper hydroxide from forming in small amounts and sinking to the bottom.

Allow the liquid to run out the bottom into the sink to fill the long glass tube completely before tightening the pinch clamp all the way and lowering the tube into the cylinder. Slide the long tube slowly down into the graduated cylinder until the tip touches the bottom. Now let the liquid flow, very slowly at first and then faster, until about 100 cm³ has gone into the graduated cylinder. Additional copper sulfate must be poured into the funnel during this filling process. When enough copper sulfate has been introduced, tighten the pinch clamp and withdraw the long tube very slowly. If this is properly done, the line of separation will be very sharp. The graduate should then be stoppered or covered with aluminum foil and placed in front of a white screen where it can be easily observed but not disturbed by your students.

Although complete mixing takes a very long time, the sharp division between the two layers disappears in a day. At the end of a week, a considerable amount of copper sulfate solution will have diffused upward into the water.

Apparatus and Materials

Copper sulfate (40 g)
Graduated cylinder (250 cm³)
Bunsen burner
Glass tube (or medicine dropper)
Pinch clamp
Funnel
Rubber or plastic tubing
Aluminum foil, or stopper to fit cylinder
Concentrated sulfuric acid (1 cm³)
Water

EXPERIMENT: GROWING SMALL CRYSTALS 9.11

Since salol (phenyl salicylate) melts at about 43°C, only moderately hot water is required to melt it. As the salol on the watch glass cools, small crystals will generally form. If they do not appear, a tiny crystal of salol can be dropped into the liquid as a seed crystal. The material on the watch glass can be melted repeatedly by placing the glass on top of a beaker of hot water. The crystals are most easily seen when the watch glass is placed on a dark, dull surface. Alcohol can be used to clean the watch glasses after the experiment.

After your students have completed this experiment, show the PSSC film, *Crystals*, with Alan Holden (25 minutes).

More information on salol and many other crystals can be found in the Science Study Series book, *Crystals and Crystal Growing*, by Alan Holden and Phylis Singer (Anchor Books, Doubleday & Company, Garden City, New York).

Answers to Questions

When a single crystal first appears, it will have a flat, rhombic shape. As it grows larger, the appearance is more like that of a stack of flat, rhombic layers of different sizes. Many crystals soon begin to grow throughout the liquid in a somewhat jumbled pattern. A given crystal stops growing when it meets the sides of other crystals. At the freezing point, the randomly arranged liquid molecules suddenly become locked together in regular, geometric patterns.

Apparatus and Materials

Magnifying lens (about 10-cm focal length)	Burner stand
	Paper towel
Watch glass	Matches
Salol (phenyl salicylate), 0.1 g	Water
Alcohol burner and fuel	Black cardboard (for dark
Beaker (100 ml)	background) if necessary

Demonstration:

A good demonstration is to show the rapid growth of crystals from a saturated solution of hypo (sodium thiosulfate, a substance used in photography).

Put one or two cubic centimeters of water into a test tube. Next add one-third to one-half of a test tube of hypo. Warm gently until all of the hypo is dissolved, and then set the tube in a beaker of cold water to cool down. When the liquid is cool, drop in a tiny bit of hypo. The resulting rapid crystallization is very dramatic!

CHAPTER 9—ANSWERS TO PROBLEMS

Sec.	Easy	Medium	Hard	Class Discussion	Home or Lab
1	1, 2†	19	8, 18	1	
2	4	3†, 5, 6, 20, 22	21	22	
3, 4	7†, 8, 10†	9, 11, 24	23	9, 11, 23	
6	13, 27, 28	12, 25, 26		12	
7		29		29	
8	31, 34	14, 15	30	14	
9	33	16, 32	35		
10	17, 36	38			37

1 What would be different about the photographs in Fig. 9.1 if (a) the bromine evaporated more rapidly and (b) there were more air in the tube?

a) The gas would become evenly distributed throughout the tube sooner than as shown in Fig. 9.1.
b) The gas would diffuse more slowly and fill the tube later than as shown in Fig. 9.1.

2† What evidence leads you to believe that atoms are in motion and not at rest?

The spreading of bromine gas when a drop of bromine evaporates, the appearance of water drops on cold window panes from a container of hot water, and the odor of ammonia spreading throughout a room are all bits of evidence suggesting that atoms are in motion.

3† If a gas is compressed until its pressure is doubled, what will happen to the density of the gas?

The density will be doubled.

4 In the sphere-gas machine, what would you expect to happen if the top disk had less mass?

The top disk would rise higher in the tube. It would also fluctuate more about its average position. (See answer to prob. 23).

5 Two bricks are placed on each of three wood dowels resting on clay as shown in Fig. A. Which dowel will sink fastest into the clay? On which is the pressure greatest?

A given load will produce a larger pressure when distributed over a smaller cross section. Thus, compared to the thicker dowels, the thinnest dowel will sink fastest in a given short time.

In all applications in the text, the cross section of any cylinder in which pressure was measured remained fixed. Therefore, we could measure the pressure in terms of the number of bricks that were placed on the piston without reference to the cross section of the cylinder. If the question of how to compare pressures when equal loads are placed on pistons of different cross section should arise, you may wish to use this problem and prob. 6. Let students argue on an intuitive basis.

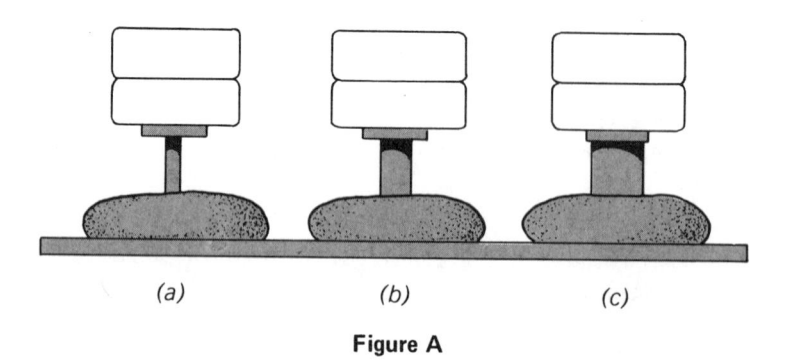

Figure A

6 Two bricks are placed on each of the pistons shown in Fig. B. In which cylinder will the pressure of the gas be greatest?

Be sure to assign and discuss prob. 5 before prob. 6. The gas in (*a*) will be subjected to the greatest pressure.

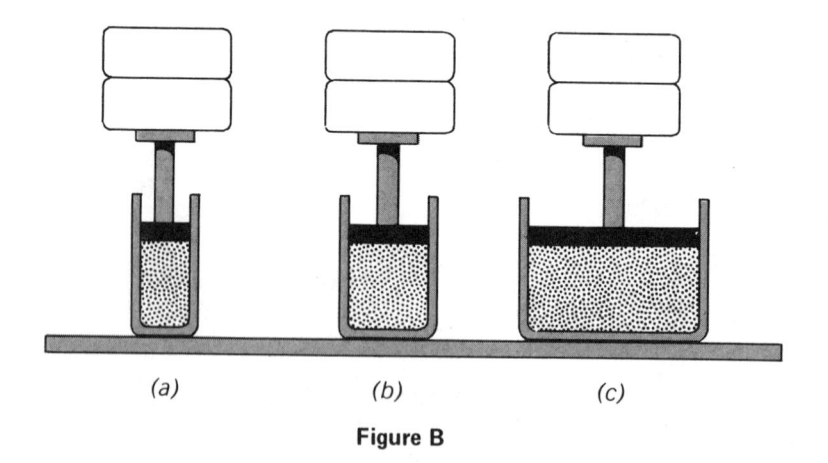

Figure B

7† From the graph in Fig. 9.11, what is the pressure when the volume is reduced (a) from V to $\frac{1}{3} V$, (b) from V to $\frac{3}{4} V$?

  a) $\underline{3P}$
  b) $\underline{4/3\ P}$

Molecular Motion 193

8 In Fig. 9.15 when the total pressure is doubled by increasing from 2.5 bricks to 5 bricks, what is the ratio of the initial volume to the final volume?

$$\frac{\text{Initial volume}}{\text{final volume}} = \frac{28.5 \text{ cm}^3}{14.7 \text{ cm}^3} = \underline{1.94}$$

9 A cylinder contains 1,000 cm³ of hydrogen at a total pressure of 1.0 atmosphere. What will be the total pressure in atmospheres if the temperature does not change and the volume is reduced to (a) 100 cm³? (b) If it is reduced to 10 cm³?

a) If the volume decreases by a factor of 1,000 cm³/100 cm³ = 10, the pressure increases by the same factor. The total pressure becomes <u>10 atmospheres.</u>
b) If the volume decreases by a factor of 1,000 cm³/10 cm³ = 100, the total pressure becomes <u>100 atmospheres.</u>

10† A cylinder contains 100 cm³ of air at a total pressure P (pressure due to piston plus atmospheric pressure). What will the total pressure become if the volume is reduced to (a) 50 cm³, (b) 10 cm³?

a) <u>2 P</u>
b) <u>10 P</u>

11 Figure 9.15 is a graph of pressure as a function of volume of air over a range of volume from about 30 cm³ to about 10 cm³. What would you predict the total pressure to be (in terms of bricks) if the volume of gas were (a) 10 cm³, (b) 70 cm³?

The graph in Fig. 9.15 may be extrapolated to get an answer to (a), but for (b) Boyle's law must be used.
a) <u>6.9 bricks</u> (by extrapolation).
b) Here the volume has increased 2.33 times 30 cm³, and so the pressure is reduced to 1/2.33 of the pressure at 30 cm³. This gives a total pressure of 2.4 bricks/2.33 or about <u>1.0 brick.</u>

12 A fast-moving tennis ball strikes a racket that is moving back, away from the ball.
a) How does the speed of the ball before it hits the racket compare with the speed after it rebounds?
b) If the piston in Fig. 9.15 is pulled up, what happens to the gas molecules when they rebound?
c) What will happen to the temperature of the gas as the piston rises?

a) The ball's speed will be less after striking the racket.
b) The gas molecules will move slower after rebounding from the moving piston.

194 Molecular Motion

c) The temperature of the gas will decrease, because the average speed of the gas molecules will be decreasing as a result of their colliding with the upward moving piston.

13 Does a bicycle pump heat up when you pump up a tire?

Yes. As you push the piston in to compress the air, molecules rebound with increased speeds from the moving piston. The average speed of the molecules increases, and therefore the temperature of the air increases. The air in turn heats up the walls of the pump.

14 If the two tubes in Fig. C contain the same liquid and the initial levels are the same, in which tube will the liquid rise higher as the temperature of the liquid in both tubes is raised equally?

The liquid in tube (a) will rise higher. Since the volumes of the containers are approximately the same, the same increase in volume will occur in both tubes, but the same volume increase will require a greater height in the narrow tube (a) than in the wider tube (b).

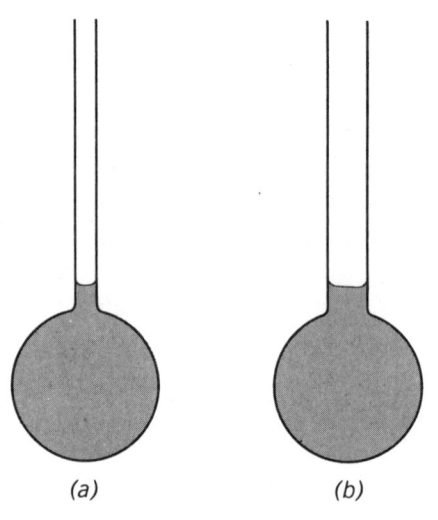

(a)   (b)

Figure C

15 One of the liquids in the apparatus in Fig. 9.20 has a volume of 33 cm³. Its tube has an internal diameter of only 0.3 cm and a volume of about 0.07 cm³ for 1 cm of its length. For a 10°C temperature rise for burner fuel, using this apparatus, the liquid rose 6.0 cm. What was the increase in volume of 1 cm³?

The increase in volume of 1 cm³ is the fractional increase in volume, this increase is, change in volume/initial volume = (6.0 × 0.07 cm³)/33 cm³ = <u>0.013</u>.

The purpose of this problem is to help the student to see that even though the rise in the narrow tubes of Fig. 9.20 is large, the change in volume per unit volume of liquid is really quite small.

16 A somewhat different apparatus for studying the thermal expansion of a tube is shown in Fig. D. Here the tube is filled with hot water, and the pointer is set at zero in a vertical position.

With this apparatus the following data were taken as the water in the tube cooled from 70.4°C to 37.8°C:

Temperature (°C)	Angle (deg)	Temperature (°C)	Angle (deg)
70.4	0	50.2	25
66.1	5	45.9	30
62.4	10	41.8	35
58.0	15	37.8	40
54.2	20		

a) Plot a graph of the angle through which the pointer turns as a function of the temperature.
b) From this form of the apparatus what information about the thermal expansion of a rod can you obtain that you were not able to get with the apparatus you actually used?

Figure D

a) See Fig. IV.
b) Evidently, equal changes in temperature produce equal changes in length throughout the extent of the temperature range observed in the experiment. The student should not assume that this holds true beyond the range measured. The "curve" may or may not remain a straight line. The student has no evidence on which to decide. In

fact, with all substances, the graph is linear (a straight line) only over a limited temperature range.

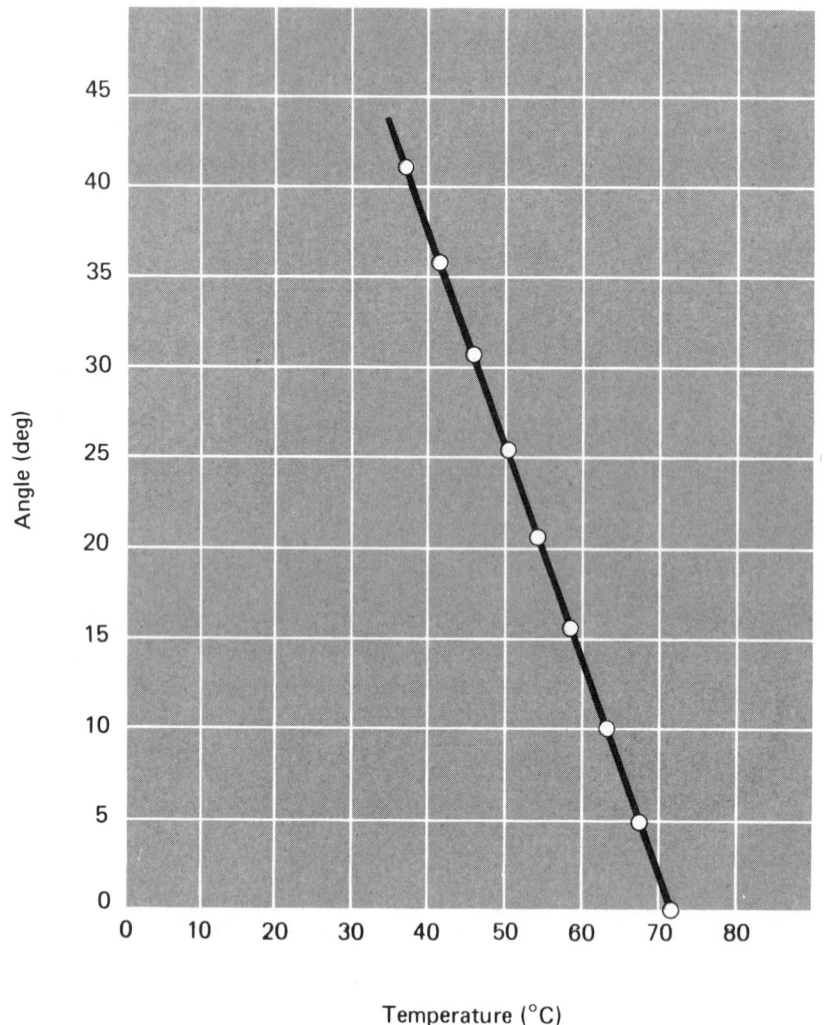

Figure IV

A graph with Temperature (°C) on the x-axis (0 to 80) and Angle (deg) on the y-axis (0 to 45), showing a linear decreasing relationship from about (35, 43) to (71, 0).

17 **The two liquids in Fig. 9.23 were kept at room temperature during the diffusion. How would you expect the pictures to differ if the experiment had been run at a higher temperature?**

Because of the increased molecular speed at higher temperatures, the diffusion would have been more advanced for the same time intervals.

18 **Bromine gas has a density of 6.5 × 10^{-3} g/cm³. Can this be the reason why the gas moves up through the air in the tube shown in Fig. 9.1?**

Bromine gas is five times as dense as air (density 1.2 × 10^{-3} g/cm³). On the basis of density the bromine gas should stay at the bottom of the tube. Difference in density cannot be the reason why the gas moves up through the tube.

19   Figure 9.2 shows that when bromine vaporizes in a vacuum, the color seems to spread immediately throughout the tube. What does this tell you about the speed of the bromine molecules?

The bromine molecules travel the length of the tube in less time than you can perceive, and so they must be moving rather fast. (Their average speed at 20°C is about 200 m/sec, or about 450 mi/hr.)

20   Would you expect an increase in pressure on the walls of the air-filled bromine tube in Fig. 9.1 as the bromine evaporates? Why?

As the bromine evaporates, the number of gas molecules in the tube increases. Since the pressure depends upon the number of gas molecules in the tube, pressure will increase as evaporation goes on.

21   Do the moving steel balls exert any pressure on the side walls of the vertical tube of the sphere-gas machine? What changes would you make in the machine to check your answer?

Since the balls hit the side wall as well as the top disk, there is a pressure on the walls. This pressure can be detected by cutting a hole in the wall and covering it with thin rubber. The rubber will be pushed out when the machine is running.

22   If you run the sphere-gas machine with a long tube and without a top disk, the density of the gas is greatest near the bottom and decreases as you go up the tube. Is there a similar effect in the atmosphere of the earth?

The atmosphere of the earth shows this very effect, with the density of the air decreasing with an increase in altitude.

23   As Figs. 9.5 and 9.6 show, the top disk does not stay at one place but moves rapidly up and down about a certain average position.
a)   How do you account for this motion?
b)   Why is such a motion not observed in the experiment described in Sec. 9.4?

a)   The irregular motion of the upper disk is caused by the irregular impact of the individual steel spheres. They are few in number, and the ratio of their mass to the mass of the disk is large enough that each individual collision of a steel sphere with the disk will cause a visible movement of the disk.
b)   In the gas experiment of Sec. 9.4, there is such a large number of collisions, and the mass of each gas molecule is so small in comparison to the mass of the platforms and bricks, that irregular collisions of individual molecules cause no noticeable variation in the position of the piston.

198 Molecular Motion

Estimating the ratio of molecular mass to piston mass and the numbers of molecules in the gas for both the sphere gas and a real gas shows how widely different the two cases are. In the experiment in Sec. 9.4,

$$\frac{\text{Mass of one oxygen molecule}}{\text{Mass of platform and brick}} \approx \frac{10^{-22} \text{ g}}{10^{3} \text{ g}} \approx 10^{-25}$$
$$\text{Number of molecules in cylinder} \approx 10^{21}$$

In the sphere-gas machine,

$$\frac{\text{Mass of one sphere}}{\text{Mass of top disk}} \approx \frac{0.03 \text{ g}}{4 \text{ g}} \approx 10^{-2}$$
$$\text{Number of steel spheres} \approx 10^{1}$$

A very small, very light mirror suspended in a partial vacuum by a very thin fiber will twist back and forth in an irregular manner as a result of bombardment by air molecules. The motion is slight. However, a beam of light reflected by the mirror can amplify the motion sufficiently for observation.

24 For three or four values of volume and pressure in Fig. 9.15 calculate the volume times the corresponding pressure. How do these products compare?

Pressure (bricks)	Volume (cm³)	P × V
3.0	24.0	72
4.0	18.2	73
5.0	14.8	74
6.0	12.2	73

The products are almost the same; $P \times V$ is a constant.

25 In what different ways can you increase the pressure of a gas?

The pressure of a gas may be increased (1) by decreasing the volume at constant temperature for a constant number of molecules, (2) by increasing the temperature at constant volume for a constant number of molecules, and (3) by adding more molecules of gas, at constant temperature and volume.

26 How is the volume of the steel-sphere gas affected by (a) motor speed, (b) number of spheres, and (c) the mass of the disk on top?

The volume of the steel-sphere gas will
a) increase with an increase in motor speed.
b) increase with an increase in the number of spheres.
c) decrease with an increase in mass of the top disk.

27 What two basic assumptions have we added to the atomic model developed in Chap. 8 in order to use it to describe the compressibility and thermal expansion of a gas?

We have assumed that the molecules of a gas are in constant motion, and that they move independently of one another.

28 Diesel engines do not ignite the fuel-air mixture with a spark from a spark plug as gasoline engines do. Instead, air in the cylinder is compressed by a piston. At maximum compression, fuel sprayed into the compressed air ignites and drives the piston back. How can you explain the ignition of the gas when there is no spark to ignite it?

As the gas in the cylinder is compressed, its molecules speed up, and the temperature of the compressed air rises high enough to ignite the fuel.

29 a) Suppose the apparatus you saw in the film loop "How Does the Thermal Expansion of Gases Compare?" were used to study the properties of neon gas. How would you expect the experimental data to compare with that shown in Fig. 9.19?
b) Suppose the apparatus were used to study the properties of a mixture of propane and carbon dioxide. How would you expect the experimental data to compare with that shown in Fig. 9.19?

a) You would expect the data points to fall on the same line as the other gases.
b) You would expect that the data points would also fall on the same line. This is a mixture of gases, as is air, which is shown on the graph.

30 a) If the liquid levels in the two tubes in Fig. E are initially the same, will they be the same or different as the temperature of the liquids is lowered the same amount?
b) If the liquids in the two tubes are different, can you directly compare their thermal expansion using these two tubes?

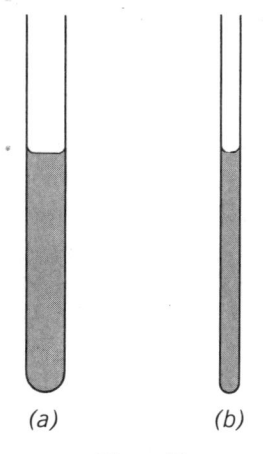

(a)   (b)

Figure E

200 Molecular Motion

a) Since the tubes are of uniform diameter, there is no amplification of the expansion. If the liquids are the same, the liquid levels will stay the same as the temperature is lowered.

b) By starting with the liquids at the same level, the change in height of the two liquids can be used to compare roughly their thermal expansion directly, if the difference in their expansion is great enough to make up for the lack of amplification.

31 The following data were obtained using an apparatus similar to that in Fig. 9.20, with water in one tube and air trapped by a drop of water in a second tube.

Temperature (°C)	Height of fluid in tube (cm)	
	Air	Water
23	0.0	0.0
25	3.9	0.4
27	7.8	0.8

What is the ratio of the change in the volume of air to the change in the volume of water for the same temperature change?

$$\frac{\text{Change in volume of air}}{\text{Change in volume of water}} = \frac{3.9}{0.4} = \underline{10.}$$

32 When heated, the glass in a thermometer bulb expands, as does the liquid that fills the bulb. Can you explain why the liquid does not go down in the tube to fill the larger volume of the bulb when the thermometer is heated?

The rise or fall of the liquid in the thermometer bulb depends on the rate of expansion of the liquid compared with that of glass. Evidently the liquid expands by a larger amount for a given temperature change than does the cavity in the bulb.

33 Some bridges have one end firmly fixed to the support at one end of the bridge while the other end of the bridge simply rests on a roller. What do you think is the reason for this type of construction?

Changes in temperature cause the length of the bridge to change because the material from which it is made will expand and contract with the rise and fall of temperature. The end resting on the roller can move easily to allow for these changes in length. If both ends of the bridge are firmly fastened, the bridge may buckle or pull apart.

34 Houses are often heated by a hot-water system in which radiators and connecting pipes are completely filled with water. Such systems sometimes are connected to an open tank above the highest radiator. What do you think is the purpose of the tank?

The volume of the water increases with an increase in temperature. The tank provides a place in which the additional volume of water produced by the expansion can be stored without being lost or damaging the system. (Since water expands more than metal pipes per °C temperature rise, if the system were full of water and completely closed, the pipes would burst.)

35 A pendulum clock (Fig. F) keeps time better if the length of the pendulum does not change much when there is a change in temperature. Would it be better to have the darker portion of the pendulum in Fig. F made of copper and the lighter portion made of aluminum, or vice versa?

The darker (inside) portions should be aluminum since they are shorter than the rest of the pendulum and must increase in length the same amount for each degree Celsius change in temperature. This will keep the net length of the pendulum constant.

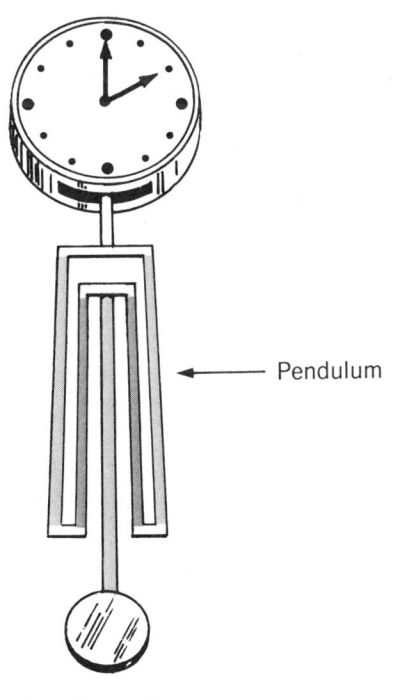

Figure F

36 If you heat one end of a short metal tube, the other end soon becomes hot. How would you account for this in terms of the atomic model?

The atoms at the hot end of the tube are moving back and forth more rapidly than those at the cooler end. The more rapidly moving atoms hit nearby atoms which are slower and speed them up. This speeding up is passed on along the length of the tube.

37 Slip a thermometer into a one-hole rubber stopper, and measure its temperature. Remove the thermometer. Holding the stopper with a pencil in one end of the hole, hammer it as rapidly and as hard as you can for a minute, and then measure the temperature. What do you observe? How do you explain it in terms of the atomic model?

The temperature of the stopper will have increased. The hammer blows cause an increase in the molecular motion of the rubber, which results in a temperature rise. (A 10°C rise is easily obtained using a No. 6 stopper.)

38 The molecules of a gas get closer together when it is compressed to very high pressures. How might you get the molecules of a gas close together without applying very high pressure to it?

We can get the molecules of a gas close together by cooling it under constant pressure—for example, in a rubber balloon or in a cylinder with a piston. In this way we can liquefy the gas without applying high pressures.

Sizes and Masses of Atoms and Molecules

10

General Comments

This chapter makes greater demands on your students' mathematical skills than any of the other chapters in the course. Although each step in some of the experiments is conceptually very simple, it takes a relatively long chain of operations with powers of 10 to get the desired results. The details should be treated only in a class where your students can go through the rather complex arithmetic without losing sight of the physical content. With students who are weak in mathematics, it may be advisable to treat the chapter lightly, carrying them through a limited number of calculations done at the chalkboard, with the main goal no more than an understanding of the method used to find the masses and sizes of atoms. For suggestions on how to do this, see comments on Secs. 10.4 through 10.6.

The PSSC film (Modern Learning Aids), *The Mass of Atoms*, parts 1 and 2, will be of great value in teaching the material in Secs. 10.4 through 10.6. Use it either as an introduction to the study of the section in the text or as a review. With a weak class you may want to do no more than show the film so that your students get the "feeling" of the experiment.

The suggested schedule for this chapter is:

Section 1-3 (two experiments, probs. 1-8; 19-21)	5 periods
Sections 4-6 (probs. 9-14; 22-30)	3 periods
Film: *The Mass of Atoms*	2 periods
Section 7 (probs. 15-18; 31-34)	3 periods
Achievement Test No. 5	2 periods
Total	15 periods

THE THICKNESS OF A THIN LAYER 10.1

This section presents the ideas that will be used in the next two experiments. Sections 10.1 through 10.3 are tightly connected, so it is

204   Sizes and Masses of Atoms and Molecules

important that each be understood before you proceed to the next.

It is very essential that your students understand that the height (or thickness) of an object whose top and bottom surfaces are parallel can be determined from area and volume measurements. In preparation for the next section you may find it advisable to ask your students to determine the thickness of the aluminum slab used in the density experiment (Sec. 3.7) from their measurements of the length, width, and mass and their calculated value for the density. They can then compare the thickness found this way with their measured value of the thickness to see that they are equal.

You may find that the concept of area needs to be reviewed with your students. A clear distinction between area and volume is essential.

10.2 EXPERIMENT: THE THICKNESS OF A THIN SHEET OF METAL

The purpose of this experiment is to prepare your students for the methods and calculations used in the next one. We are not primarily concerned with an accurate determination of the thickness of a piece of aluminum foil.

The Experiment

You may wish to hand out aluminum foil of one thickness to some groups and of a different thickness to the others. In a post-laboratory discussion, you can then make a histogram of the class results on the chalkboard, similar to the one shown in Fig. I.* The grouping of results around two different values clearly shows that the class was using metal of two different thicknesses. The aluminum-foil thickness will, of course, vary from class to class, depending on the source, but usually it is between 0.001 cm and 0.003 cm. Therefore, in answer to the questions in the text, we can say that the atoms in the foil cannot be larger in diameter than this value, but they could be one-half or one-third as thick, and so on.

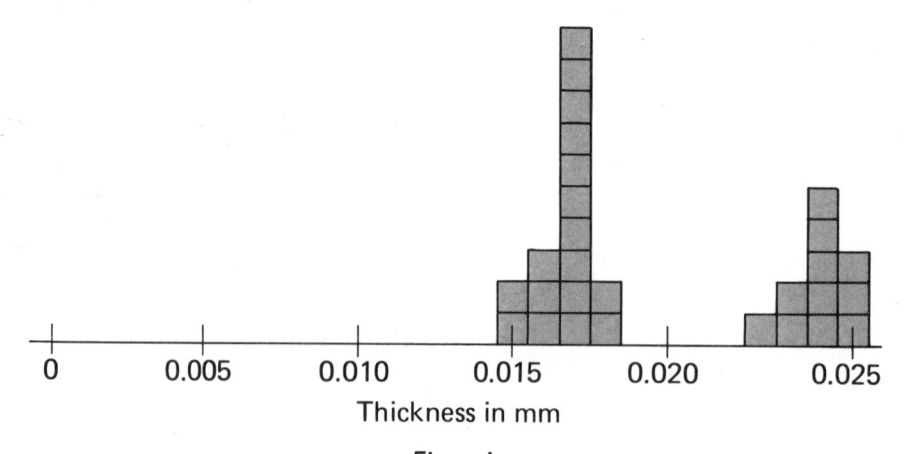

Figure I

*The value for the thickness of 4.0×10^{-3} cm from the sample data was clearly in error and was not included in the histogram.

Sample Data

Length (cm)	Width (cm)	Area (cm²)	Mass (g)	Volume (cm³)	Thickness (cm)
7.5	7.6	57	0.38	0.14	2.5×10^{-3}
7.5	7.5	56	0.23	0.085	1.5
7.7	7.7	59	0.39	0.14	2.4
7.5	7.6	57	0.26	0.096	1.7
7.5	7.5	56	0.37	0.14	2.4
7.6	7.6	58	0.27	0.10	1.7
7.4	7.4	55	0.6	0.22	4.0
7.6	7.4	56	0.27	0.10	1.8
7.6	7.6	58	0.39	0.14	2.5
7.4	7.4	55	0.25	0.093	1.7
7.7	7.5	58	0.38	0.14	2.4
7.7	7.5	58	0.24	0.090	1.5
7.5	7.7	58	0.35	0.13	2.2
7.7	7.5	58	0.25	0.093	1.6
7.6	7.4	56	0.25	0.093	1.7
7.6	7.5	57	0.36	0.13	2.3
7.6	7.5	58	0.27	0.10	1.7
7.7	7.7	59	0.38	0.14	2.4
7.5	7.5	56	0.25	0.093	1.7
7.5	7.5	56	0.37	0.14	2.4
7.3	7.3	53	0.24	0.089	1.7
7.5	7.5	56	0.26	0.096	1.7
7.7	7.4	57	0.24	0.089	1.6
7.7	7.6	59	0.40	0.15	2.5
7.5	7.5	56	0.27	0.10	1.8
7.7	7.6	59	0.37	0.14	2.3
7.5	7.7	58	0.26	0.096	1.7
7.6	7.5	57	0.39	0.14	2.5
7.7	7.6	59	0.26	0.096	1.7
7.7	7.5	58	0.25	0.093	1.6

Apparatus and Materials

Balance
Centimeter rule
Aluminum foil, regular and heavy duty; pieces about 8 cm × 8 cm

EXPERIMENT: THE SIZE AND MASS OF AN OLEIC ACID MOLECULE 10.3

This experiment has two parts. The direct part yields the thickness of the layer and can be found to one significant digit, using the entire class data.

The calculation of the number of molecules in the second part is more difficult than calculating the thickness and involves a more difficult line of reasoning. It may be omitted for weaker classes without destroying the logic of the chapter.

Some of your students may object that they cannot measure the size of the oil drop and the diameter of the oil patch accurately and that, therefore, their results are not to be trusted. However, you can point out that without doing this experiment they would know only that molecules are too small to be seen (prob. 1); they would have no idea whether they are smaller than 10^{-4} cm, 10^{-14} cm, or even 10^{-20} cm!

Problems 2 through 6 help prepare your students for some of the steps encountered in this experiment. You may wish to do prob. 4 as a demonstration to aid their understanding further, checking the result with a ruler or other direct measuring device.

During the pre-lab, be sure to discuss the problem of estimating the size of the drop of oleic acid on the wire. After a small drop has been placed on the wire, have several students estimate its diameter, using a centimeter rule and a magnifying glass, and calculate its volume. There will be considerable disagreement in the diameter estimates, but the difference should not be greater than 0.03 cm. The volume differences, however, will be much larger because the cube of the diameter is involved. In doing the experiment, both students in a pair should make several measurements of the diameter of the drop.

Some students may argue that the volume of the drop should be calculated from the formula for the volume of the sphere. Actually, the drop is not spherical, as your students can see when they measure it. Furthermore, you can show them that the error in measuring the size of the drop is so large that we can justify results in which only one digit is significant.

You can also have your students estimate the area of the oil film by assuming it to be a square with one side equal to the diameter of the "circle." In fact, making both assumptions will result in a smaller error than assuming only that the drop is cubical. Making both assumptions will not seriously affect the results but will simplify the calculations for students who have difficulty with the matematics.

To compare the errors in making one and in making both assumptions, you can proceed as follows:

Assumptions	Volume or Area
Spherical drop	$V = \frac{4}{3}\pi \left(\frac{\text{diam. sphere}}{2}\right)^3$
	$\approx \frac{(\text{diam. sphere})^3}{2}$
Cubical drop	$V = (\text{diam. sphere})^3$
Circular layer	$A = \pi \left(\frac{\text{diam. circle}}{2}\right)^2$
	$\approx \frac{3(\text{diam. circle})^2}{4}$
Square layer	$A = (\text{diam. circle})^2$

Assumptions	Thickness		Error
Spherical drop, circular layer	$\dfrac{\frac{(\text{diam. phrere})^3}{2}}{\frac{3(\text{diam. circle})^2}{4}}$	$= \dfrac{2\,(\text{diam. sphere})^3}{3\,(\text{diam. circle})^2}$	—

Cubical drop, circular layer	$\dfrac{(\text{diam. sphere})^3}{\dfrac{3(\text{diam. circle})^2}{4}} = \dfrac{4\,(\text{diam. sphere})^3}{3\,(\text{diam. circle})^2}$	100%
Cubical drop, square area	$\dfrac{(\text{diam. sphere})^3}{(\text{diam. circle})^2}$	33%

Instead of going through the preceding, rather involved, argument, you may have students good at arithmetic who will volunteer to calculate and compare the size of the oleic acid molecule using (1) only the cube assumption and (2) both assumptions, with the size obtained by making neither assumption.

The Experiment

Trays with a diameter of at least 15 inches are required. PSSC ripple tanks or large cafeteria trays are satisfactory. It is easier to see the edge of the oleic acid film if the ripple tank is used on a black table. Metal trays can be painted black or lined with dark-colored plastic material. A tray can also be made by cutting off a large cardboard carton near the bottom and lining this bottom section with aluminum foil or dark-colored plastic. Do not allow your students to carry trays full of water to and from the sink, because they will be sure to spill the water. The trays should be emptied by tilting a corner into a bucket, or by siphoning into a bucket with a length of rubber tubing.

The best powder to use is lycopodium powder (the spores of the common ground pine). Shaking it on the water surface from a small cloth bag makes it easy to control the amount. Your students will tend to use too much powder; very little is needed. If too much powder is put on the water, the film will not reach its full size and a single layer of molecules will not result.

Supply your students with pure oleic acid in small bottles and a 4-inch piece of No. 32 tinned copper wire. This can be obtained by taking apart a length of No. 24 stranded copper wire containing seven strands. The No. 32 wire is 0.2 mm thick.

With a narrow V, the drop size will have an average diameter of about 0.2 mm to 0.7 mm. (A 0.7-mm drop may produce a film too large for the tray.)

Warn your students not to dip the wire too deeply into the oil. (If they do, oil will cling to the wire above the drop that is formed in the V.) If this occurs, have them wash the wire off in burner fuel. Ask your students what effect dipping the wire too deeply would have on the results of the experiment.

The "circles" formed by the oil drop are generally quite circular and about 14 to 25 cm in diameter.

Sample Data

Individual estimates of the size of a given drop may vary by about 0.03 cm, but with practice the variation usually is less.

208 Sizes and Masses of Atoms and Molecules

Average diameter of droplet (mm)	Diameter of thin film (cm)	Thickness of layer (cm)
0.5	20	4×10^{-7}
0.3	14	2
0.3	21	0.8
0.3	12	3
0.4	16	3
0.5	19	5
0.5	21	4
0.2	16	0.4
0.2	13	0.6
0.4	15	4

All of the values in the table are of the same order of magnitude, 10^{-7} cm.

Although a histogram of the class results (Fig. II) shows the data scattered over a wide range, be sure your students realize that before they did the experiment the only answer they could give to the question "How big is a molecule?" was "Smaller than I can see." Now they can answer, in the case of oleic acid, "Assuming the layer was one molecule thick, the molecules are about 10^{-7} cm high."

Thickness of layer in cm

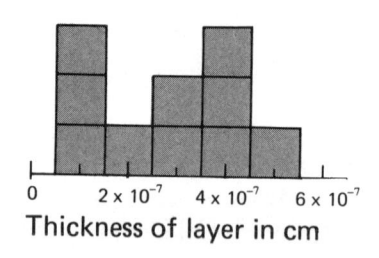

Figure II

Answers to Questions

The length of the side of the "cubical" drop is about 0.3 mm.

If the film is only roughly circular, one can visualize a circle that has an area equal to that of the film and record its diameter.

The diameter is about 10 to 20 cm.

The layer is about 3×10^{-7} cm thick.

$$\text{Area of base of mol.} = \left(\frac{\text{height of mol.}}{10}\right)^2$$

$$\approx \left(\frac{3 \times 10^{-7} \text{ cm}}{10}\right)^2 = 9 \times 10^{-16} \text{ cm}^2$$

$$\text{No. of mol.} = \frac{\text{area of layer}}{\text{area of base of mol.}}$$

$$\approx \frac{3 \times 10^2}{9 \times 10^{-16}} = 3 \times 10^{17}$$

The mass of the droplet was

$$\text{Mass} = \text{vol.} \times \text{density}$$
$$= 10^{-4} \text{ cm}^3 \times 1 \text{ g/cm}^3 = 10^{-4} \text{ g}$$

The mass of a single molecule of oleic acid is

$$\text{Mass of mol.} = \frac{\text{mass of drop}}{\text{no. of mol.}}$$

$$\approx \frac{10^{-4} \text{ g}}{3 \times 10^{17}} = 3 \times 10^{-22} \text{ g}$$

Apparatus and Materials

Large tray (15 in. × 15 in.)
Pure oleic acid (1 cm^3)
Magnifying lens
Meter stick for measuring diameter of circle
Centimeter scale (small), for drop-size estimates
Wire (4 in. length of tinned copper No. 32 gauge)
Burner fuel
Lycopodium powder (1 cm^3 in small cloth bag)
Water
Plastic bucket
Paper towel

THE MASS OF HELIUM ATOMS 10.4
THE MASS OF POLONIUM ATOMS 10.5
THE SIZE OF POLONIUM AND HELIUM ATOMS 10.6

The details of the radioactive processes and the method of detecting radioactivity are treated here as a "black box." It is not advisable, during the class discussion, to attempt to explain the mechanism of a radiation counter.

For the sake of simplicity, we have limited our discussion to the production of helium atoms in the decay of polonium, without mentioning alpha particles. Actually, it is the alpha particles that are detected by the counter. They become helium atoms, however, as soon as they come in contact with any solid material, and under the conditions of the experiment described in these sections, this takes place in less than 10^{-10} second. The term "alpha particle" dates back to the time when the nature of these particles was not yet understood, and it was not known that they were nuclei of helium atoms. Since students have no evidence from this course for the existence of atomic nuclei, you should not get involved in a discussion of the nature of alpha particles.

If your class is not very proficient in understanding arithmetic involving powers of 10, you can simplify the treatment of Secs. 10.4 through 10.6 in the following way: Emphasize only the basic method used in the experiment; that is, the mass of particles too small to mass directly can be found by measuring the mass of a large sample and counting the number of particles in the sample by some means (radioactive disintegration counts in this case). The total mass divided by the number of particles then gives the mass of a single particle.

$$\text{Mass of particle} = \frac{\text{total mass of sample}}{\text{no. of particles}}$$

You can omit the mathematical details of dilution and counting, and simply use the final total count and the mass of helium obtained from its volume and density to arrive at the mass of the helium atom. You can, if necessary, omit the details of the method of determining the mass of polonium that disintegrated (discussed in Sec. 10.5) and use only the final result to calculate the mass of the polonium atom.

The film, *The Mass of Atoms*, showing Mr. Hertz and Mr. Brewer actually performing the experiment described in the text, was made specially for this course. The final results given in the film are not the same as those given in the text, since each result represents a different trial. The film, which comes in two reels of about 20 minutes each, should be shown on two successive days.

The following gives the principal steps taken to determine the masses of the helium and polonium atoms. After these steps is a flow sheet that summarizes the theory and calculations involved in determining the masses and volumes of polonium and helium atoms.

Mass of Helium Atom

1. A sample of polonium is sealed in an evacuated tube and left for three weeks.
2. The tip of the quartz tube is broken under water and the volume of trapped helium is measured.
3. The sample of polonium is diluted in nitric acid.
4. The rate of disintegration of the sample of polonium is determined with a radiation counter, and from this the number of helium atoms produced in three weeks is found.
5. From the mass of the sample of helium and the number of atoms in the sample, the mass of a single helium atom is calculated.

Mass of Polonium Atom

1. Mass the polonium sample. This was, of necessity, done at the beginning of the experiment with helium.
2. Dilute the polonium (same as Step 3 for mass of helium).
3. Determine number of polonium atoms that decayed in 3 weeks (same as Step 4 for mass of helium).
4. Determine mass of polonium that decayed in 3 weeks (from Step 1 and decay curve of polonium).
5. From the mass of polonium that decayed (Step 4) and the number of polonium atoms that decayed (Step 3), determine the mass of one polonium atom.

Mass of Helium Atom

Steps in Calculation	Numerical Calculations
I *Volume of Helium:* Area of cross section × Length of tube	8.1×10^{-3} cm² × 5.0 cm = 4.0×10^{-2} cm³
II *Mass of Helium:* Volume × Density of helium	4.0×10^{-2} cm³ × 1.7×10^{-4} g/cm³ = 6.8×10^{-6} g
III *Fraction of Polonium Sample Counted:* $a \times b \times c$ a is fraction taken of first solution b is fraction taken of second solution c is fraction of b placed on counter	$\dfrac{1}{1{,}000} \times \dfrac{1}{100} \times \dfrac{1}{1{,}000}$ = $10^{-3} \times 10^{-2} \times 10^{-3} = 10^{-8}$
IV *Number of Helium Atoms Formed in 3 Weeks:* $\dfrac{(\text{counts/min}) \times 2^{\dagger} \times (\text{time in min})}{(\text{Fraction of sample counted})}$	2.2×10^5 counts/min × $\dfrac{2 \times 3 \times 10^4 \text{ min}}{10^{-8}}$ = 1.3×10^{18} atoms
V *Mass of Single Helium Atom:* $\dfrac{\text{Mass of helium}}{\text{Number of helium atoms}}$	$\dfrac{6.8 \times 10^{-6} \text{ g}}{1.3 \times 10^{18} \text{ atoms}} = 5.2 \times 10^{-24}$ g/atom

†Only half the particles, those flying upward, are counted.

Mass of Polonium Atom

Steps in Calculation	Numerical Calculations
I *Number of Polonium Atoms Decayed in 3 Weeks:* Number of helium atoms formed in 3 weeks	1.3×10^{18} atoms
II *Mass of Polonium Decayed:* Fraction decayed × Mass of sample	$\frac{1}{10} \times 4.5 \times 10^{-3}$ g = 4.5×10^{-4} g
III *Mass of Polonium Atom:* $\dfrac{\text{Mass of polonium decayed}}{\text{Number of polonium atoms decayed}}$	$\dfrac{4.5 \times 10^{-4} \text{ g}}{1.3 \times 10^{18} \text{ atoms}} = 3.5 \times 10^{-22}$ g/atom

Volume of Helium Atom

I *Volume of Helium Produced (when liquid):* $\dfrac{\text{Mass of helium}}{\text{Density of liquid helium}}$	$\dfrac{6.8 \times 10^{-6} \text{ g}}{0.15 \text{ g/cm}^3} = 4.5 \times 10^{-5}$ cm³
II *Volume of Single Helium Atom:* $\dfrac{\text{Volume of helium (when liquid)}}{\text{Number of helium atoms}}$	$\dfrac{4.5 \times 10^{-5} \text{ cm}^3}{1.3 \times 10^{18} \text{ atoms}} = 3.5 \times 10^{-23}$ cm³/atom

Volume of Polonium Atom

I *Volume of Polonium That Decayed:* $\dfrac{\text{Mass of polonium decayed}}{\text{Density of polonium}}$	$\dfrac{4.5 \times 10^{-4} \text{ g}}{9.3 \text{ g/cm}^3} = 4.8 \times 10^{-5} \text{ cm}^3$
II *Volume of Single Polonium Atom:* $\dfrac{\text{Volume of polonium decayed}}{\text{Number of polonium atoms decayed}}$	$\dfrac{4.8 \times 10^{-5} \text{ cm}^3}{1.3 \times 10^{18} \text{ atoms}} = 3.7 \times 10^{-23} \text{ cm}^3/\text{atom}$

10.7 ATOMIC MASSES AND MOLECULAR FORMULAS

This section ties together the knowledge of atomic masses with the laws of constant and multiple proportions studied in Chaps. 6 and 8. This combination enables us to find ratios of the numbers of different atoms in compounds and to determine the simplest molecular formulas. It should be kept in mind that except in gases and some liquids and solids, these formulas do not imply the existence of molecules as clusters of atoms. It might be helpful to review Sec. 8.9 again at this time.

The masses of the atoms in Table 10.1 were not determined by radioactive decay. The most accurate data come from mass spectrograph measurements.

Sizes and Masses of Atoms and Molecules   213

CHAPTER 10—ANSWERS TO PROBLEMS

Sec.	Easy	Medium	Hard	Class Discussion	Home or Lab
1	2†	1	19		
2	5†	3†, 6†	4, 20	4	
3	7, 8†	21			
4	10†, 12†, 26	9, 11, 22	25	11	
5	28	13, 14, 27	23	13	
6	29		24, 30		
7	16, 31, 33, 34	15†, 17, 18, 32			

1 How do you know that there are more than 200 molecules in a spoonful of water?

If water is sprayed out of an atomizer onto a smooth table top, you can see drops that are smaller than a millimeter in diameter. (Fog droplets are even smaller.)

If we assume water molecules are as large as cubes 1 mm on a side, then a teaspoonful of water (about 5 cm³) would contain

$$\frac{5 \times 10^3 \text{ mm}^3}{1 \text{ mm}^3} = 5 \times 10^3 \text{ molecules}$$

You can demonstrate that there are more than 200 molecules in a spoonful of water by placing the water on a table top, adding a tiny drop of detergent, and then spreading the liquid with your fingers to cover as large an area as possible. You can easily spread it over a square whose area is greater than 4×10^3 cm². Since the volume is the thickness times the area, the thickness of the layer will be

$$\text{Thickness} = \frac{\text{Volume}}{\text{Area}} = \frac{5 \text{ cm}^3}{4 \times 10^3 \text{ cm}^2} \approx 10^{-3} \text{ cm}$$

If the molecules are assumed to be cubes whose side is equal to the thickness of the layer, then the number of molecules in the teaspoonful of water is

$$\text{Number} = \frac{\text{Total volume}}{\text{Volume of one molecule}} = \frac{5 \text{ cm}^3}{(10^{-3} \text{ cm})^3} = 5 \times 10^9$$

This problem is a good introduction to Expt. 10.3, The Size and Mass of an Oleic Acid Molecule.

2† A cube has a side of 2×10^{-7} cm. What is its volume?

8×10^{-21} cm³

3† In a small-boat harbor a careless sailor dumps overboard a quart (about 1,000 cm³) of diesel oil. If we assume that this will spread out evenly over the surface of the water to a thickness of 10^{-4} cm, what area will be covered with oil?

10^7 cm²

214 Sizes and Masses of Atoms and Molecules

4 Spherical lead shot are poured into a square tray 10 cm on a side until they completely cover the bottom. The shot are poured from the tray into a graduated cylinder, which they fill to the 20-cm^3 mark?
   a) What is the diameter of a single shot?
   b) How many shot were in the tray?
   c) If the 20 cm^3 of shot weighed 130 g, what was the mass of a single shot?

a) The diameter of a single shot would be the same as the thickness of the layer. Assuming the shot pack directly over one another in the graduate, the thickness is

$$\frac{\text{Total volume}}{\text{Area}} = \frac{20 \text{ cm}^3}{10 \text{ cm} \times 10 \text{ cm}} = \underline{0.2 \text{ cm}}$$

b) Number of shot = $(10 \text{ cm}/0.2 \text{ cm})^2 = \underline{2{,}500}$

c) Mass of single shot = $\dfrac{\text{Total mass}}{\text{Number of shot}}$

$$= \frac{130 \text{ g}}{2{,}500} = \underline{5.2 \times 10^{-2} \text{ g}}$$

This problem prepares students for the experiment in Sec. 10.3, since it is a direct analogue of that experiment, using lead shot in place of oleic acid molecules.

5† A tiny drop of mercury has a volume of 1.0×10^{-3} cm^3. The density of mercury is about 14 g/cm^3. What is the mass of the drop?

$\underline{1.4 \times 10^{-2} \text{ g}}$

6† A goldsmith takes 19.3 g of gold and hammers it until he has a thin sheet of foil 100 cm in length and 100 cm in width. The density of gold is 19.3 g/cm^3.
   a) What is the volume of gold?
   b) What is the area of the gold sheet?
   c) What is the thickness of the gold sheet?

a) $\underline{1.0 \text{ cm}^3}$
b) $\underline{1.0 \times 10^4 \text{ cm}^2}$
c) $\underline{1.0 \times 10^{-4} \text{ cm}}$

7 If 3×10^{-5} cm^3 of pure oleic acid forms an oil film with an area of 150 cm^2, how thick is the film?

$$\text{Thickness of film} = \frac{\text{Volume}}{\text{Area}} = \frac{3 \times 10^{-5} \text{ cm}^3}{1.5 \times 10^2 \text{ cm}^2} = \underline{2 \times 10^{-7} \text{ cm}}$$

8† How many oleic acid molecules occupy 1 cm³ if the volume of one oleic acid molecule is 10^{-23} cm³? Assume that the molecules are 10 times as high as they are wide and that there is no empty space between them.

10^{23}

9 If the mass of an atom of an element is 5.0×10^{-23} g, how many atoms are there in 1g of that element?

$$\text{Number of atoms} = \frac{\text{Total mass}}{\text{Mass of one atom}} = \frac{1 \text{ g}}{5.0 \times 10^{-23} \text{ g/atom}}$$

$$= 2 \times 10^{22} \text{ atoms}$$

10† A cylindrical tube with a cross-sectional area of 2 cm² and a height of 50 cm is filled with hydrogen. What is the volume of the hydrogen in the tube?

100 cm³

11 Write a brief summary of the steps followed in finding the mass of helium atoms by radioactive decay.

The steps in the experiment are as follows:
1. Measure the volume of helium produced by polonium in a closed tube in a given time, and calculate the mass of helium from the density.
2. Dissolve in nitric acid the polonium that was in the closed tube.
3. Count the number of disintegrations in one minute of a known fraction of the polonium.
4. From (3) and the time in (1), calculate the total number of polonium atoms that disintegrated in the tube, which equals the number of helium atoms produced.
5. The mass of a single helium atom is then found by dividing the mass of helium (Step 1) by the number of helium atoms determined in Step 4.

12† In the experiment on the mass of helium, how many lead atoms were formed? What assumptions are made to get this number?

1.3×10^{18}. We assume that the disintegration of one polonium atom produced one helium atom and one lead atom.

13 Write a brief summary of the steps followed in finding the mass of polonium atoms by radioactive decay.

216 Sizes and Masses of Atoms and Molecules

1. Mass the polonium sample and proceed with the determination of the number of helium atoms produced by the sample during a given time. This is also the number of polonium atoms that have decayed.
2. From the decay graph of polonium, find what fraction of the mass of polonium decayed during this experiment.
3. Use (1) and (2) to find the mass that decayed. Then divide by the number of polonium atoms that decayed to find the mass of a single atom.

14 In Expt. 10.3 you determined the mass of an oleic acid molecule. How many atoms would this molecule contain if the mass of each atom equaled (a) the mass of a helium atom or (b) the mass of a polonium atom?

a) Since the mass of an oleic acid molecule is about 3×10^{-22} g, the number of atoms equal in mass to a helium atom is

$$\frac{3 \times 10^{-22} \text{ g}}{5.2 \times 10^{-24} \text{ g/atom}} = \underline{60 \text{ atoms}}$$

b) The number of atoms (equal in mass to a polonium atom) is

$$\frac{3 \times 10^{-22} \text{ g}}{3.5 \times 10^{-22} \text{ g/atom}} = \underline{1 \text{ atom}}$$

15† Assuming the smallest mass you can measure on your balance is 0.005 g, what is the smallest number of copper atoms you could mass on your balance?

About 5×10^{19} atoms

16 Use Table 10.1 to find the mass in unified atomic mass units of a molecule of phosphorus pentoxide (P_2O_5).

The mass of one phosphorus atom is 31.0 u, and the mass of one oxygen atom is 16.0 u. Therefore, the mass of one molecule of phosphorus pentoxide is 2×31.0 u $+ 5 \times 16.0$ u $= \underline{142.0 \text{ u}}$

17 In Sec. 6.2 you found that the mass ratio of hydrogen to oxygen in water was 0.13 = ⅛. Using Table 10.1 find the simplest formula for water.

The ratio $\frac{\text{mass of one hydrogen atom}}{\text{mass of one oxygen atom}} = \frac{1 \text{ u}}{16 \text{ u}}$. Since the observed ratio in water $\frac{\text{mass of hydrogen}}{\text{mass of oxygen}}$ is twice this value, there must be two hydrogen atoms in water for every oxygen atom. Thus the simplest formula is $\underline{H_2O}$.

18 In Table 8.3, the ratio of the mass of chlorine to iron in iron dichloride is given as 127/100. Use information from Table 10.1 to determine the simplest formula for iron dichloride.

From Table 10.1, the mass of one chlorine atom is 35.5 u and that of iron is 55.8 u. The simplest assumption is that one chlorine atom combines with one iron atom, which gives a mass ratio of chlorine to iron of 35.5/55.8 = 0.64, about half of the ratio that is given. Combining two atoms of chlorine with one atom of iron gives a mass ratio (2 × 35.5)/55.8 = 1.27, which equals 127/100. Thus the simplest formula is $\underline{FeCl_2}$.

19 A rectangular object 3.0 cm × 4.0 cm × 5.0 cm is made of many tiny cubes, each 1.0×10^{-2} cm on a side. How many cubes does the object contain?

The volume of the tiny cubes is 1.0×10^{-2} cm × 1.0×10^{-2} cm × 1.0×10^{-2} cm = 1.0×10^{-6} cm³. The volume of the rectangular object is 3.0 cm × 4.0 cm × 5.0 cm = 60 cm³. Therefore,
Total number of cubes = 60 cm³/1.0×10^{-6} cm³ = $\underline{6 \times 10^7}$

20 The diameter of a tennis ball is about 0.07 m, and the dimensions of a tennis court are 15 m × 30 m. How many tennis balls will be required to cover the court?

If you assume the tennis balls are arranged in a square and not in a hexagonal or close-packed arrangement, each ball will effectively occupy an area of $(0.07 \text{ m})^2$. The total number of balls that are required to cover the court would be

$$\frac{\text{Area of court}}{\text{Area of each ball}} = \frac{15 \text{ m} \times 30 \text{ m}}{(0.07 \text{ m})^2} = \underline{9.2 \times 10^4}$$

21 If the molecules of the oleic acid monolayer you made could be placed end to end in a line, about how long would it be?

The length of an oleic acid molecule is about 3×10^{-7} cm. Therefore, since the drop of acid used contained about 3×10^{17} molecules, placing them end to end would give a length of $3 \times 10^{-7} \times 3 \times 10^{17} = \underline{9 \times 10^{10} \text{ cm}} \approx \underline{600{,}000 \text{ mi.}}$ They would stretch to the moon and back!

22 If a 10^{-3}-g sample of radium gives a count of 4×10^7 counts/min, how much radium would give 100 counts/min?

A sample having a mass of

$$\left(\frac{100}{4 \times 10^7}\right) \times 10^{-3} \text{ g} = \underline{2.5 \times 10^{-9} \text{ g}}$$

218 Sizes and Masses of Atoms and Molecules

23 Suppose you buy a 2-kg bag of dried beans and you find it has been contaminated with small stones. How would you go about finding the approximate number of the stones in the bag without separating all the stones in the bag from the beans? What assumptions have you made?

Take a small sample of the contents of the bag. Pick out the stones and the beans and mass them separately; then find the mass of a single stone and a single bean. Find the value of the

ratio: $\dfrac{\text{mass of stones in the sample}}{\text{mass of sample}}$. Multiply this ratio by 2 kg to

get the mass of stones in the bag. Divide this mass by the mass of one stone to find the number of stones in the bag.

You have assumed that (a) the sample is representative of the mixture of stones and beans in the bag; and (b) the stones all have the same mass, and the beans all have the same mass.

24 A cubic millimeter (10^{-3} cm^3) of blood is found to contain about 5×10^6 red blood cells. An adult human body contains about 5×10^3 cm^3 of blood. About how many red blood cells are there in an adult human body?

The number of cubic millimeters in the blood of an adult human body is about $\dfrac{5 \times 10^3 \text{ cm}^3}{10^{-3} \text{ cm}^3} = 5 \times 10^6$. Each of these contains 5×10^6 cells, and so there must be about $5 \times 10^6 \times 5 \times 10^6 = \underline{2.5 \times 10^{13} \text{ cells}}$ in an adult human body.

25 Some polonium is dissolved in 1,000 cm^3 of nitric acid, and a 0.01-cm^3 sample of the solution is counted. Then the number of disintegrations per minute is found to be 3×10^3. How many disintegrations per minute occurred in the original solution?

The total number of disintegrations per minute in the original solution is

$$\dfrac{10^3 \text{ cm}^3}{10^{-2} \text{ cm}^3} \times 3 \times 10^3 \text{ dis./min} = \underline{3 \times 10^8 \text{ dis./min}}$$

26 If 10^{18} atoms of polonium disintegrate to produce lead and 10^{-5} g of helium, what is the mass of a helium atom?

If each atom of polonium produces one helium atom, the mass of one helium atom is

$$\dfrac{\text{Mass of helium}}{\text{Number of helium atoms}} = \dfrac{10^{-5}}{10^{18}} = \underline{10^{-23} \text{ g}}$$

27 Use Fig. 10.9 to determine the fraction of a polonium sample that remains after 45 days. What fraction of the polonium decayed during this time?

From the decay curve, 0.8 of the polonium remains after 45 days, and thus, 1.0 − 0.8 = 0.2 of the polonium decayed during this time.

28 a) What fraction of a sample of pure polonium will decay in 100 days? (See Fig. 10.9.)
b) If a counter initially records 5×10^4 counts/min for the sample, what would you expect it to record after 100 days?

a) 1.0 − 0.6 = 0.4 will decay in 100 days.
b) From the graph, 0.6 of the original sample remains after 100 days. Since the counts per minute are proportional to the mass of the sample remaining, the counter will read $0.6 \times (5 \times 10^4$ counts per minute$) = 3.0 \times 10^4$ counts per minute.

29 What would you get for the volume of one atom of helium if you calculated it from the equation below?

$$\text{Volume of atom} = \frac{\text{volume of gas sample}}{\text{number of atoms in gas sample}}$$

The volume of helium collected during the experiment was 4.0×10^{-2} cm^3, and the number of atoms in this volume was 1.3×10^{18}. Therefore, the volume of a single atom would be

$$\text{Volume of atom} = \frac{4.0 \times 10^{-2} \text{ cm}^3}{1.3 \times 10^{18} \text{ atoms}} = 3.1 \times 10^{-20} \text{ cm}^3$$

This is, of course, too large by a factor 10^3, since in the volume of the gas sample the atoms were far apart compared to their size.

30 A penny is about 1 mm thick. About how many layers of copper atoms does it contain?

If we assume that copper atoms have a diameter of 3×10^{-8} cm, the number of layers of copper atoms is

$$\frac{\text{Thickness of a penny}}{\text{Thickness of a copper atom}} = \frac{10^{-1} \text{ cm}}{3 \times 10^{-8} \text{ cm}}, \text{ or about } 3 \times 10^6$$

31 What is the ratio of the mass of carbon to the mass of oxygen in carbon dioxide (CO_2)?

From Table 10.1, the mass of one carbon atom is 12.0 u and that of one oxygen atom is 16.0 u. Each molecule of carbon dioxide contains one atom of carbon and two atoms of oxygen, and so the ratio

$$\frac{\text{Mass of carbon}}{\text{Mass of oxygen}} = \frac{12.0}{2.0 \times 16.0} = \underline{0.375}$$

32 Using the atomic masses of copper and chlorine in Table 10.1 return to the data you obtained from the experiment with two chlorides of copper in Chap. 8, and determine the simplest formulas for the two compounds.

From the histogram in the *Teacher's Guide* for Expt. 8.7, we see that 112 g of chlorine combine with 100 g of copper in the brown chloride, and 55 g of chlorine combine with 100 g of copper in the green chloride. The mass ratios for the two compounds are therefore:

	Brown Chloride	Green Chloride
$\frac{\text{Mass of Cl (g)}}{\text{Mass of Cu (g)}}$	$\frac{112}{100} = 1.12$	$\frac{55}{100} = 0.55$

If we assume that the simplest formula for one of the compounds is CuCl, then the ratio of the atomic masses is

$$\frac{\text{Atomic mass Cl}}{\text{Atomic mass Cu}} = \frac{35.5}{63.5} = 0.56.$$

Since this is very close to the ratio of the masses of these elements as found in the experiment with the green compound, the green chloride has the simplest formula, $\underline{\text{CuCl}}$.

Because the mass ratio found in the experiment with the brown chloride is 2 × 0.56, the ratio of the atomic masses in the brown compound is (2 × 35.5)/63.5 = 1.12, and its simplest formula is therefore $\underline{\text{CuCl}_2}$.

33 Could a sample containing only sodium and oxygen have a molecular formula NaO if the ratio

$$\frac{\text{Mass sodium}}{\text{Mass oxygen}}$$

in the compound is 2.9? Explain. (See Table 10.1.)

<u>No</u>. From Table 10.1, the mass of a sodium atom is 23.0 u and that of an oxygen atom is 16.0 u. The ratio of mass sodium/mass oxygen for NaO would be 23.0/16.0 = 1.44. Thus, the sample is not NaO. Since 2 × 1.44 = 2.9, the simplest formula is Na_2O.

34 What ratio of combining masses would you expect in a compound containing (a) one atom of lead to every atom of oxygen or (b) one atom of lead to every two atoms of oxygen?

a) $\dfrac{\text{Mass of 1 lead atom}}{\text{Mass of 1 oxygen atom}} = \dfrac{207}{16.0} = \underline{12.9}$

b) $\dfrac{\text{Mass of 1 lead atom}}{\text{Mass of 2 oxygen atoms}} = \dfrac{207}{32.0} = \underline{6.47}$

Appendix

Listed below are the approximate quantities (class of 24) and the minimum purity specifications for the chemicals needed in the IPS course. When replacing chemical supplies care should be taken to insure that the purity and standards of the chemicals are the same or better than listed here. Failure to do so will cause poor results in the experiments and in a few cases could be hazardous.

Chemical	1	2	3	4	5	6	7	8	9	10
Alcohol, Denatured Ethyl, 95% – 1 gal.	x		x	x	x	x		x	x	x
Alcohol, Methanol – 1 pt. (optional)			x	x						
Alcohol, Isopropyl, USP 99% – 1 pt.					x					
Alka-Seltzer – 2 dozen tablets			x	x						
Aluminum Foil, Heavy – 1 roll										x
Aluminum Foil, Regular – 1 roll						x			x	x
Aluminum Strips, 0.05 × 1 × 10 cm – 30								x		
Ammonia Solution, Household – 1 pt.				x						
Boiling Chips – ½ kg										
Citric Acid, Gran., Hydrous, UPS – 100 g					x					
Copper Dust, Purified Electrolytic – 100 g						x				
Copper, Fine Gran., Reagent – 100 g		x								
Cupric Sulfate, Powder, Tech. – 100 g		x							x	
Cupric Chloride, Anhyd. Powder, Reagent – 100 g								x		
Cuprous Chloride, Anhyd. Powder, Reagent – 100 g								x		
Epsom Salt-Magnesium Sulfate, USP – 100 g			x	x						
Glycerin (Glycerol), USP – 1 pt.	x		x		x	x			x	
Hydrochloric Acid, Conc. Reagent – 1 pt.								x	x	
Ink, Black Washable – 1 pt.					x					
Lead Nitrate, Crystals, Tech. – 100 g			x							
Limewater, USP – 1 pt.				x						
Magnesium Carbonate, block, USP – 30 g				x						
Magnesium Ribbon – 1 roll				x						
Naphalene, Flakes – ½ kg			x							
Paradichlorobenzene Crystals – ½ kg			x							
Potassium Nitrate, Gran., USP – ½ kg						x				
Potassium Dichromate, Gran., USP – 150 g						x				
Sand, washed – 1 kg		x								
Salol (Phenyl Salicylate), Crystals, USP – 100 g									x	
Sodium Carbonate, USP – 1 kg						x				
Sodium Chlorate, Crystals, Tech – ½ kg						x				
Sodium Chloride, Fine, USP – ½ kg					x	x				
Sodium Iodide, Gran., USP – ½ kg	x									
Sodium Nitrate, Gran., Purified – 100 g					x	x				
Sugar, Gran. – ½ kg					x	x				
Sulfur, Powder, USP – 100 g		x			x					
Sulfuric Acid, Conc. Reagent – 1 pt.						x			x	
Zinc, Commercial, 0.05 × 1 cm × 1 cm – 120								x	x	